Georg E. Schäfer

# HISTORY OF COMPUTER SCIENCE

2nd Edition

The contents of this book were compiled and edited with great care. All the same, the author and the publisher accept no liability and provide no guarantee that there will be no errors in this book. Whenever the reader finds errors or misleading information and reports this to the author, the information will be used to improve the next edition. If a copyright or trade mark symbol is not attached to product names or other information, this does not indicate that there is no copyright. Contributions and comments are welcome especially from 'veterans' who worked for many years with Information and Communication Technology.

Please mail to georg-ernst.schaefer@web.de.

This book is dedicated to Hilde, Julia, Florian and Paul, Leon and Jona. May it inspire them and all other readers.

Impressum:

Bibliografische Information der Deutschen Nationalbibliothek:
Die Deutsche Nationalbibliothek verzeichnet diese Publikation in der Deutschen Nationalbibliografie; detaillierte bibliografische Daten sind im Internet über dnb.dnb.de abrufbar.



Herstellung und Verlag:
BoD – Books on Demand, Norderstedt
ISBN 9783751999267

9 783751 999267

# Table of Contents

# INTRODUCTION

Computer Science is no longer limited to the delivery and operation of services to simplify our daily life. Ten years ago we were happy with services like navigation of our cars through places where we have never been, the automatic update of our mobile devices and simple possibilities to communicate with friends and relatives using social networks. Today computing power accompanies us day and night. Avatars considered in 2000 a dream of a faraway future now work for us imperceptibly. Computing in one of the many instances find new proofs for century old mathematical problems. Social networks find mates from our early childhood by automatic face recognition no matter where they live and clouds provide automatic synchronization of documents on the globe. Analysing big databases with medical data, computers propose cures your medical expert might not have considered. Some smartphone apps or health bands around our arm tell us more about the state of our body than intensive traditional medical check-ups. But hate messages, fake news, cyber attacks and political misuse e.g. to influence national elections are a fact, too.

> **Fuzzy borders in timelines**
> Throughout this book, information about the date of inventions and discoveries is provided. We recall what most experts know. The first version of a product or specification is rarely a mature tool for practical use. A product, released in 1970, needs a sales force to explain it to the customers, delivery to customize and install it, field engineers to maintain and update it and teachers to explain the potentials. So the reader should add three or five more years to the release date to imagine the computing reality of the past.

Substitutions of human reasoning by computer algorithms occur in today's world in a perpetually faster way. Exponentially growing knowledge, financial crisis, COVID-19 therapies, rapid establishment of a mobile phone infrastructure in Africa, matchless economic rise of China, and military and cultural conflicts now coexist on a global level. New biological viruses, found in one part of the world, are neutralized a few days or weeks later with universal knowledge from other parts of the world. Open and fast processing of information provides the only possibility to manage this diversity in a prosperous, peaceful world with the freedom of speech, research and of stability. Computer Science is at the heart of this transformation.

In spite of such concussions, computers – except quantum computes - still work in a step by step calculation mode. Their simple arithmetic operations are not much different from the early instructions of the central processing units. What efforts were taken? What concepts were used?

In spite of all achievements, there seem to be white islands. All computer intelligence installed in the world could not tell what had happened in March 2014 to flight MH 370. This flight of a Boeing-777 disappeared from the radar screens and got lost in the Indian Ocean with 390 people in it. There are good reasons to understand the relative position between humans and computers. What are computers doing? Where can we trust this new virtual world and where is suspicion mandatory? In mid-2020, people desperately hope that computers could find medication and vaccine for COVID-19.

Computing empowers billions of people, rich and poor. Computing was used successfully in settling conflicts, building new companies and creating jobs. Computing is not positive by itself. Many organizations and politicians felt painful consequences of social media systems and revelations of documents. Some social changes were successful because interested groups used advanced computing technology, like video servers and smartphones. Computing needs to be organized and shaped to serve us. If we do that, and there is no alternative, computers will never take control of our mind and world.

The language of today's global information and knowledge society is English. The rise of ubiquitous use of Information and Communication Technology (ICT) is connected to the world wide use of English. That is why this book is also written in English. The history of English informs us about a variety of English languages like American, British, Indian, African, European or Australian English.[1] Like the different versions of English language the ways to the information and knowledge society are and will be individual paths on a mainstream progress.

Tomorrow, robots might help us at home, when we are sick, or at work. More powerful information and picture processing will help us to become more productive and thus cope with demographic challenges. Some readers may smile when they read this, thinking that this will never happen. They are in respectable company. Even the researchers and inventors of information and communication systems regularly thought so in the 1950s. They should have been insiders at the time. Nevertheless they did not see the technologies and methods emerge that shaped today's world, forty to sixty years later.

One or two big computers will be enough for the whole world, a CEO of a leading computer company said, when that "big computer" had less computational power than a cheap cell phone in the 1990s. Are we more aware of the technologies and methods that pioneers develop today for the next 40 years?

A closer look, how transformation happened in about 100 years from 1910 till 2020, is time well spent. History uncovers timelines of industrial, financial and sociological progress. Extrapolating these trends, history empowers us to navigate our understanding of the future.

Born in 1950, the author studied Mathematics, Physics and Computer Science. During his professional career, he worked with the computer systems of all major manufacturers and in different application areas. Starting as system analyst, he later supported the sales division, implemented new projects and infrastructures. When he joined the government, he helped to build up the privacy commissioner's agency and later coordinated information and communication technologies (ICT) on European level as well as on Local, Federal and State level in Germany for two decades. As an employee of a American company and later as customer, the author got a deep insight into the laboratories and production sites of the leading American and European ICT companies.

Much of this book's information derives from the author's personal documents and his own experience. The emphasis is on general computing and not on the microprocessors in machines like laboratory equipment. For readers with interest in this area, we recommend Stephen G. Nash's "History of Scientific Computing".[2] Our goal is to provide an easy to understand, fascinating report on the history of computing for all people interested in computing, no matter what their professional training was.

"A historian who merely reads about a field without getting deeply involved and making significant contributions himself is apt to be a mere voyeur. He will not understand the act of creation, and hence he will miss the essence of what we want …".[3] The author hopes the reader gets more profit from this book than from a heartless report of an ignorant voyeur.

Most insiders of Computer Science saw their work utterly misunderstood by some. The author remembers a discussion with jurists in his team. They seriously asked him to develop a police computer that can only function according to the law. The response, that a bread knife can always be used to kill people, opened their perspective. Technology conquers new and unexpected applications. In 1948, Howard Aiken, designer of the MARK I calculator, saw only a commercial market for five or six of such calculators.[4]

We hope that this easy to read presentation of some basic ideas, that brought global knowledge in our homes and mobile phones, will allow readers to assess computing comparable to other evolutionary technologies. If humans live in an incomprehensible environment, they lose orientation and motivation to design life and society. Computing needs to be explained in an easy way, just like biology, electricity, gas networks and transportation with motor vehicles, bikes and solar vehicles. There is a further objective: The reader should see that computing is built on very simple ideas. Much innovation is possible. We only just made the first step in computing.

The overall objective is to reduce the myths many get in mind when dealing with computers. Some feel, computers are mathematics and much more of the things they never ‘really understood’ in school. There must be some magic about it. Myths should be instanced to assess the realistic implications.[5]

We hope many readers see in the timelines the dominant trends of progress in the areas covered by this history book. The trends enforced each other and brought, again and again, new functionality to the market. Piling up material goods, provides only a linear increase in a pile of goods. Who piles up information gets an exponential increase of knowledge. How that? Adding “-“ and “>” produces an arrow “→”. If we change the position of these two items we get “↓” or “↑” with different meanings. This meaning was not transmitted with “-“ and “>”. Art, poetry and architecture work on this “growth” basis. Computing and knowledge management work likewise. Considerations about the nature of knowledge and information are part of Information Theory.[6] Since around 2000, we understand content syndication and the way it empowers our systems. Mashup web applications combine geo-referenced data with other content like cyclone warnings, traffic information, burglary incidents or house prices. The technical aspects are reduced to one of many considerations, although still of great importance, in this context.

Law, contracts, other regulation and politics protect, organize and shape modern technology and how it is applied. There are ideas like freeware, community built open source systems, and proprietary, quality assured software as well as net neutrality as one basis of social and commercial development of multibillion euro and dollar markets. The privacy laws rule the way we process personal data with computers. Copyright legislation says who can use programs and design concepts. A history of technology is not complete without looking at these subjects.

The author asked veterans from computer technology to provide further details. These experts felt uneasy about possible contributions humbly referring to their narrow view on the technology. A few former colleagues provided valuable information and support. The author especially thanks his wife Hilde for sustainable encouragement and help and his brother Fred Schäfer for a lot of invaluable information and constant motivation.

Some papers about the history of Artificial Intelligence, Neural Networks etc. tell us that these concepts were first introduced in the 1940s and 1950s. No doubt, researchers of this time speculated and projected computing power to learn by itself, to solve existential problems for mankind and to "think". All this was not what today we consider project work delivering results in a financially defined time frame. This history book reduces its report to the events when precise algorithms or solutions became available.

# 1. INTENTIONS OF THIS HISTORY OF COMPUTER SCIENCE

Although we live in the information society, where computers and microprocessor are the basis of prosperity and everyone's daily companions, we do not have a picture of the history of computer science as a whole. Many books and other documents were published on specific historical aspects of computer science.[7] Researchers and inventors told their lives and explained their achievements.[8] [9] [10] [11] A free collection "Computer Science Bibliographies" is available in the Internet.[12] There is no integrated documentation of the development of computer science, customer's needs and legal solutions, applications and the mathematical and logical foundations. "The history of technology in general is not well documented. In fact, it is hardly documented at all."[13]

> **THINK is the slogan of IBM since 1924.** Charles Flint integrated three little companies June 16, 1911, and Thomas Watson Sr. joined him in 1914. Howard Aiken built the first calculator Harvard Mark I for IBM, a non-binary but programmable calculator. The commercial success of IBM began with the IBM 603 Electronic Multiplier. Later IBM proved to be particularly successful with its mainframes using for decades the same basic code and instruction set architecture. IBM set quality standards with its architectures (like e.g. System Network Architecture SNA, System Application Architecture SAA, with several cryptographic architectures). IBM's PC operating system OS/2 had important advantages over DOS / Windows for the computing environments of big customers. IBM was the first company that invented "computing on demand" which it developed to cloud computing. IBM is still a leading technology provider, research company and an important service provider e.g. with the Smarter Planet Concept. (Kevin Maney, Steve Hamm, Jeffrey M.O'Brien, Im Dienst der Welt, IBM Press – Pearson Plc, 2011)

A book outlining a 70-year timeline of ideas around computers – technological, mathematical, organizational, financial and legal - that shaped society will undoubtedly provide a lot of insight. Information and Communication technology are either the reason for transformation or a catalyst. When the veterinarians found out that the mating of animals according to pre-calculated strategies based on some algorithms (using animal characteristics) improves food supply, European politicians passed a law introducing animal registers. Soon other continents, like the U.S.A., followed. Robots, e.g. for medical hip operations, improved the health systems worldwide. "Just-in-time" logistics revolutionized industry. These are only a few examples that demonstrate the power of computing based transformations. Where are we going? What transformations come next? How can we foresee and manage them? These are vital questions. One answer results from an extrapolation of the history of computing.

We decided to try to fill that gap in library shelves of the history books. For the author, it was a pleasure to compose it. If someday others will, under the same objective, write another book and highlight other aspects, we could get further help to understand ICT history. To further advance computer science it is helpful to look back. Many ideas, we might feel being new, were already drawn up by others.

Gottfried Wilhelm Leibniz (1646 – 1716) considered clear human thinking as a kind of calculation. God – he thought – calculated with his more powerful means the world. Therefore, he invented a suitable language and algebra (called *characteristica universalis* ) with many other concepts like e.g. binary digits. He built a calculation machine (*supra hominem*, "better than mankind"), based on the decimal system, in 1673 and demonstrated it at the Royal Society in London.[14] From today's point of view, Leibniz abilities stayed much behind his expectations. A long row of similar bold and far thinking personalities worked together, some developing methods, others developing devices and hardware, and some extremely industrious and intelligent people coded millions of lines. The history of computing science is not only a history of technology but also a history of motivation, tenacity and industry.

Industry indeed was necessary, because Kurt Gödel disappointed in 1931 philosophers and mathematicians by establishing that the correctness of models and theories cannot be proved. Error-free computer systems and theories are impossible to be built, except for trivial and simple situations. So students of computer science learned that the absence of errors can never be proved. A lot of time has to be spent to find the obvious errors with elaborated testing for hours, day and night. Students of law schools learned that computer users have to pay providers of software and computer systems to get information about the errors and deficiencies of products. Such legal constructs are not possible for "conventional" products like cars, buildings or dish washers. Normally, guarantees, that come with the products, allow us to give malfunctioning merchandise back to those who produced or sold it. As computer systems are never faultless, computer and software manufacturers provide guarantees only through costly maintenance. Maintenance costs of much of modern commercial software are currently about 20% per year of the price of the original licence, in Europe.

Gödels findings about the extremely limited possibilities of mathematicians to prove a theorem still hold today. We are never sure if a set of axioms of practical importance is really non-contradictory and non-redundant. What mathematicians find even more difficult to prove is the existence of the objects they defined with their axioms. But scientists are working hard to find automated provers. Several methods were found and are currently applied. Computers proved mathematical theorems, first in the 1960s. The hope is that they could check mathematical proofs of 100 pages or more. In 2012/2013 experts from all over the world met in Princeton, NJ, and discussed the current state of the art. Many theorems could already be proved with computers and new insight was found. But the state of the art in 2014 would best be described by: "Although automation is an exciting and ambitious goal, there is little realistic hope of having automated provers routinely prove assertions with real mathematical depth."[15] Writing software could be seen as creating a mathematical theory on the application area of this software. Automatic provers for mathematical theories could test large software systems. The pessimistic result of the mathematicians working in this filed is an important reason why software quality is restricted and will be restricted for the next decades.

The thinking of Leibniz and Abu Dscha'far Muhammad ibn Musa al-Chwarizmi (780 B.C. – 850 B.C.), whose name is the origin of the word "algorithm", lead to the idea of a calculation, which is done in a sequential and step-by-step approach. Parallel computing or associative computing was not seriously investigated at that time. Alan Turing formalized that concept with his Turing Machine. Its calculation steps were thought to be on a tape illustrating in that way a sequential processing mode. When Johann Neumann (John von Neumann) looked at the problem from an engineering point of view he envisaged the computer as a device with data and instructions being in one single memory.[16] [17] This single memory was structured in cells numbered from 0 to any integer which looked technologically feasible and realistic at the time. Thus the machine architecture, as we still know it, was drawn up.

.

We here use the concept of pseudo code to demonstrate the concept. The integrated circuit had still to be invented, and the microprocessor was not yet on the market:

| Number of Memory Cell | Content |
|---|---|
| 0 | Start program e.g. for the calculation of 5! |
| 1 | Read starting number from cell number 1000 |
| 2 | Do one or more operations with the starting number |
| 3 | Go back to memory cell 2 if the endpoint is not yet reached |
| 4 | .... |
| 1000 | starting number, let's say: 5 |

These ideas may look rather simple, like many ideas in Computer Science. However, Computer Science proofs from the beginning that considerable theoretical efforts are necessary to achieve small practical success. At a time when technicians were still experimenting with mechanical solutions, with slow tubes, practically no memory and poor quality, the software engineers and scientists produced visionary concepts. The following table shows some milestones[18]:

| **Year** | **Software / Methods** | **Hardware** |
|---|---|---|
| 1941 | Alan Turing presents the concept of a universal machine (Turing machine). | |
| 1943-1944 | John Mauchly and J. Presper Eckert (members of the University of Pennsylvania) build with vacuum tubes the Electronic Numerical Integraor and Calculator (ENIAC). | The ENIAC (1942) was mostly programmed by plugboards not by languages. Rotating magnetic memory follows the early insufficient thermal and mechanical concepts. |
| 1947 | Flowcharts developed (Goldstine, John v. Neumann) | |
| 1945-1950 | Machine-level instructions, Code generating algorithm | Most machines were of bit serial designs and devices with mixed serial/parallel design |

| | | |
|---|---|---|
| 1952 | First compilers | UNIVAC I had a main memory of 1000 words of 12 characters each. Input/output was done externally on separate devices. |
| 1956 | Mark I DB IBM 702, early database manager, FORTRAN developed | The machines could be viewed as tabulating calculators that justified first ideas of more systematic and powerful digital computers. |

Advances in other scientific fields were necessary to achieve technological breakthrough. When the microprocessor was developed in a further step, programming languages with data structures, re-usability and code sharing became a necessity. It was then that far looking scientists and engineers found an Operating System (OS) a convenient way to use the fast processors. The OS allowed first parallel running programs and later using the same computer in a time sharing mode.

Fig. 1: Storage Media from the 1970's till today

Like today, people were thinking that the computing power of their time would be more than anyone needed. That was and is short sighted. After the time sharing systems were marketed, e.g. the Multi Virtual System (MVS) of IBM, engineers found that the path length (calculated in terms of 'number of instructions') to control the terminals was too long for such an expensive machine. Little and cheap microprocessors in the terminals could make the whole system cheaper. Although IBM did an excellent job developing computers in that direction, they did not clearly enough see that semi-intelligent terminals are the natural father of the PC and the 3-tier computing architecture.

> **Konrad Zuse** (June 22, 1910 – December 18, 1995) began in 1935 in Berlin to construct a computer in his parents' home. In 1938, he presented his concept based on tubes and conventional lamps. He told it could achieve up to 10.000 clock cycles per second. His audience considered his imagination running wild. Zuse constructed prototype computers Z1, Z2 and finally his Z3. In May 1941, the world's first Turing-complete computer Z3 became operational with binary 22 Bit operations, floating point arithmetic, 64 word storage, and controlled by tapes with 8 channels. Output was displayed on lamps and the speed was 3 sec. for multiplication, division and square root operations. In 1945 during World War II, Zuse developed his Z4 and a high level programming language "Plankalkül". His company ZUSE KG came into trouble in 1960, as in the U.S. many people understood the potential of computers much better than in Europe. (Konrad Zuse, Der Computer Mein Lebenswerk, Springer Verlag Berlin, 1984)

The PC came, and a hierarchy (or 'farm') of servers challenged the mainframe and – in some people's minds – even the super computer. The three tier concept worked well with the technology at hand at the time for database technology, application and network servers.

The process of challenge and response is not only a concept of cryptography. It is the way of technological progress. The market now offers smartphones and car media systems. Soon we may look through Internet-driven spectacles, from Google or other companies.

Fig. 2: Why Main Memory was so expensive!

Although the path of development was, is and will forever be long and difficult ('stony'), computer science is rapidly developing, not only as a science itself but as a motor for the development of all areas of today's society.

We now live in the information and knowledge society and computers. This became evident when in 1999 enterprises and governments were anxious whether computer systems process properly the change from 31.12.1999 to 01.01.2000. Many old and badly maintained application programs had hundreds of lines of program code with hard coded "19xx" date fields. These application programs could not be used after 1st January 2000. In 1999, a migration to other provider's programs was compulsory. Shares of the efficient ICT companies listed on the share market of the "New Economy" rose to unprecedented heights. In 2000 - 2001, the Dotcom bubble of the share market collapsed. Although many inefficient companies left the market, and the share prices of ICT companies listed considerably lower, the share market recovered soon on a stable level. Another example for the speed of changes and the extraordinary level of transparency inherent to the information and knowledge society provides the global financial crisis 2007 to 2008. The data and the problems were soon understood. Of course, the political and social protagonists needed and – looking especially at the Euro zone – still need time to reorganize society and economy. This is one reason for the Open Government movement that started prominently with Barak Obamas presidency. One could even go back to the fall of the Soviet Union with e.g. the re-unification of Germany. It is a fact that, after 1989, the treaty for the reunification of Germany could only be completed and unanimously approved by all parliaments and parties concerned within less than one year after the fall of the Berlin Wall on November 9, 1998, because electronic Office Automation Systems and other ICT applications were used.

What strikes everyone, who begins to study Computer Science as an amazing area of human thinking, is the gap between the incredible achievements (like artificial intelligence or big data analysis) and the difficulties to produce error-free software. Handling errors under a framework of limited budget, insufficient time and tough objectives is a substantial occupation of a project manager.

In 1978, a computer with a central processing unit (CPU), able to do about 1 million calculations per second, cost about 1 million Euros. Today we use "Ultrabooks" or smartphones, slate or tablet computers with several processing units working in parallel, each 1000 times faster than those earlier CPUs. This computing power is on sale today for a fraction of the prize. 30 years ago we were happy to have a hundred or so electric motors to help us vacuum cleaning or starting our cars. Today, one person uses even more electric gear plus a hundred or more artificial intelligence units ('avatars') that help us manage daily life: Cell phone, television, translator, DVD device and navigation unit in the car. Hundreds of Apps and App-management tools are daily companions. People trust these ubiquitous intelligent avatars, although there is no way to prove their correctness.

The question is how mankind could create this huge virtual world with such a bad methodology at hand. This is one question that transforms history of computer science to a thrilling subject.

The pioneers, that built this world, were geniuses of technology, organization, and economy. Take e.g. Gene Amdahl. When the internationally active IBM Corporation built with thousands of well-educated engineers the biggest and fastest computers, he left IBM and built even faster ones that consumed less energy with only a few collaborators. He not only built them, but he also sold them. He had found out that the bigger a team becomes it needs more inter-team communication and documentation. Thus, teams tend to become less productive and less innovative the larger they become. Tomas J. Watson remembers that IBM had to learn that lesson when system software for the /360 computer family was far behind schedule: "We learned the hard way one of the great secrets of computer engineering: throwing people at a software project is not the way to speed it up." He recalls a statement of Fred Brooks, who once wrote according to Thomar J. Watson, "the bearing of a child takes nine months, no matter how many women are assigned."[19]

There are only a few documentations of the history of computer science, and these are mostly reduced to specific, small areas of computing. We are beginning to draw the whole picture focussing on the relation of mathematical models, hardware development progress, software development concepts and languages. The reason is that most scientists are specialized and feel uneasy to report about developments in other areas. To write this compendium we need the courage to be incomplete or throw a spotlight on a subject. We try to give reference so every reader can dig deeper in the subjects mentioned and find out more details.

## 2. THE HALL OF FAME OF THE PIONEERS OF COMPUTER SCIENCE

Computer Science consists of an enormous amount of inventions. One invention can be seen to be built in a hierarchical system on the base provided by other innovations. E.g., the Apps of a smartphone are built on top of an operation system like Android, which is built with one or more programming languages. These programming languages use data structures that constitute a science in itself.

Innovation never ends. Data structures, taken as an example, could be improved. Instead of using every day a new protocol file to store system information, a wraparound file system would need less maintenance and disk space. ICT needs pioneers with better ideas, even for the concepts and methods we are used to for years. Other sciences proceed the same way: Physics describes the world as it is understood at time X1, X2, X3, … with continuously improved notations (models). Computer science is to a high degree an invented world open for improvements in many directions. Maybe, like in many mathematical areas, there is already someone sitting at a kitchen table writing down a revolutionary new concept!

People are relevant, with their ideas and their fierce intention to bring these ideas into a virtual world. To honor the pioneers of Computer Science we now show a table with some of the most distinguished engineers and thinkers (as representatives of the many). To present a complete table, we expect to need some more years and sustainable help from the ICT community.

When does this timeline of computer science begin? Some think, we should begin when the first algorithm was formulated by Abu Dscha'far Muhammad ibn Musa al-Chwarizmi (in short: al-Chwarizmi) in Bagdad. His name is the root of the word "algorithm". We might begin the history of computing with the Boolean Logic, invented by George Boole (1815-1864). Both inventions[20] reduced the complexity of human's general reasoning enormously and were of greatest importance for the development of computers. To see, that rational thinking could be valued by only two logical attributes, namely 'true' and 'false', lead with the invention of binary digits by Leibniz to the basic concept of the 'von Neumann computer architecture'. Later logicians found out that n-valued logic (with logical values like e.g. "true", "may be", "false") gives no new insight and can be reduced to the Boolean algebra. Thus, a well-founded basis for the universal application of computers was given.

To reduce this timeline to a practical dimension, it is focused on the primary pioneers of Computer Science and explains along this timeline the leading technical inventions. The choice of the pioneers is rather arbitrary, although those pioneers were chosen that stand for the development of a new paradigm of their time. Readers who look for a complete description of pioneers (professional experience, bibliography) should refer to exciting books like e.g. the book of J.A.N. Lee, Computer Pioneers,[21] of Dennis E. Shasha and Cathy A. Lazere, Out of Their Minds[22].

| Time | Name | Short Homage |
|---|---|---|
| 1697 | Gottfried Wilhelm Leibniz | Endowed with an extraordinary ability for structured and abstract thinking, Leibniz created integrated concepts especially in philosophy (system of monads), jurisdiction and mathematics (e.g. probability theory and infinitesimal calculus). He had a good eye for optimizations. With his mathematical understanding of the world, he argued that God created the best possible world for us. - The Leibniz wheel was the first commercially produced calculator. - If we had lived in the 17th and 18th century, many other thoughts than computing would have been on our priority list. Remember, that in 1609 Galileo Galilei built his first telescope and Johannes Kepler described the revolutionary concept of planetary motion. In 1687 Isaac Newton published his *Principia Mathematica,* in 1785 Charles Coulomb introduced the inverse-square law for electrically charged objects. In 1800 Alessandro Volta presented his concept of a battery. |
| 1843 | Charles Babbage, Ada Lovelace | Ada Lovelace (her full name was: Augusta Ada King Byron, Countess of Lovelace) worked as a team member with Charles Babbage.[23] She was the only legitimate child of Lord Byron. As women were not allowed to enter libraries, her husband copied scientific articles for her. Ada Lovelace completed the description of Babbage's Analytical Engine, and she first programmed and tested this program with a computer using as an example the calculation of the Bernoulli numbers. - Konrad |

| | | |
|---|---|---|
| | | Zuse recalls that the achievements of Charles Babbage were mostly forgotten. Babagge built his machines with pure mechanic parts. Conditional calculation steps were already part of his scheme. The U.S. patent office drew Konrad Zuse's attention to Babbage's patents after World War II. [24] It must have been a bitter fate for Babbage to have logically working concepts for a calculator in mind without the technical means at hand. |
| 1889 | Hermann Hollerith | Hollerith worked for the American Census Bureau. With the concepts of Joseph Marie Jacquard and Charles Babbage, who controlled their machines with punched programs, he built a special machine to create a "punched photo" of every American. On July 6, 1911, Hollerith sold his company for 2.312.100 US-$ to Charles Flint and it thus became part of IBM. IBM later standardized the punch card to 80 columns. Programs and data were stored on cards. |
| 1912-1954 | Alan Turing | Alan Turing was a British mathematician, logician, crypto specialist and code breaker. He invented the Turing Machine in order to formalise algorithms by the concept of a computing mechanism that works on a sequence of instructions (on a tape) with unlimited memory. These instructions can be repeated. The tape can be moved back and forward. He thus presented the basics of the procedural processing of the Central Processing Units of the computers. Together with Alonzo Church's lambda-calculus as a basis for syntax and semantic of computer programming languages and Gödels recursive theory, the basic concepts of the procedural programming languages were worked out. Later object oriented and graphical programing was invented. |
| 1941 | Konrad Zuse | Konrad Zuse (see also box above) designed and built the first programmable digital computer. The resources of post World War II Europe were totally insufficient. In the U.S. many scientists and companies had the courage to invest in this technology. The U.S. Academic Program was growing fast and fuelling the emerging industry. Academic Research and Development in computer science and related fields rose in the U.S. from about 400 Mill. US-$ in 1973 to nearly 5 Bill. US-$ in 1999.[25] |

| | | |
|---|---|---|
| 1944 | Howard Hathaway Aiken | Howard Aiken constructed the first IBM Automatic Sequence Controlled Calculator (ASCC), Harvard Mark I, which was completed in February, 1944. As the electronic used for television sets was produced for a mass market, computers were expected to follow that design principle. Working without binary modules, it was nevertheless programmable with punched tapes. The Atanasoff-Berry-Computer, completed in the USA in 1941, was binary coded and nonetheless not programmable. Even the ENIAC in 1948 was not working with binary operations, but it was the first computer without any mechanical parts. |
| 1945 | Neumann, John von. | John von Neumann was an exceptionally productive mathematician and physicist. Among many other discoveries and inventions, he helped to construct computers. He postulated the 'von Neuman Architecture of Computers', which is still used: Computer programs and their data are stored in one computer memory and are transmitted between the different modules of the computer on a common cable system which is called a „bus". |
| 1947 | William Hewlett and David Packard | In 1939, the two engineers started their enterprise in Packard's garage. They worked on analogue electronic technology like oscillators. In 1947 HP incorporated. The semiconductor technology was first used in the 1960's. |
| 1956 | Fritz-Rudolf Güntsch | With the leadership of Fritz-Rudolf Güntsch, Telefunken Computer released in 1970 the fastest computer produced at the time in Europe, the TR 440. In his PhD thesis for the Technische Universität Berlin Fritz-Rudolf Güntsch invented the virtual memory and demonstrated its functions on a computer with 10 drums. The virtual memory was a decisive step towards the general purpose computer as the extremely fast CPU had no memory big enough and cheap enough to run programs efficiently. |

Fig. 3: List of Instructions and Assembler Programm of TR 440

| | | |
|---|---|---|
| 1958 | Jack St. Claire Kilby, Robert N. Noyce | In 2000, Jack Kilby received the Nobel prize for inventing the Integrated Circuit (IC). The Bell Telephone Laboratories developed the transistor in 1947. Jack Kilby had the vision that computing successfully deals with the "Tyranny of the large numbers" if all computing elements would be out of the same material. At the time (1958) he began to work with Texas Instruments (TI). He presented the first very simple IC in September 1958.[26] TI used the IC in 1966 for calculators. |
| 1962 | Donald Knuth | Donald Knuth began in 1962 to write a book that should have informed about everything a programmer needs to know. Scientific progress proved to be much faster than he could write. His first book appeared in 1968. More volumes followed. His work helped more than a generation of programmers. |
| 1966 | Joseph Weizenbaum | Many remember Joseph Weizenbaum as a passionate critic of the overestimation of computers. Especially the founders of Artificial Intelligence were in the mid-1960's convinced they could make within ten or twenty years computers capable of doing all human work. As opposing view Joseph Weizenbaum invented a pseudo-psychiatrist named ELIZA. It used simple rhetoric patterns but appeared to be able of understanding the computer user and conducting meaningful conversations. His contributions were helpful for a realistic understanding of the abilities of computers and for the development of the Information Society. His professional achievement was the definition of an early version of a programming language with several connections to Artificial Intelligence. |
| 1968 | Gordon Moore | In 1968 Gordon Moore was one of the founders of Intel Corporation. Before, he developed IC, memory, and processor chips. He is known for his prediction called Moore's Law: "The number of transistors of a chip seems to double every year." A popular translation said that computer processors become twice as fast and half as expensive as the comparable chips four years ago. Computer Scientists soon thought in the categories of "Technology Generations". One generation was considered to take three to four years. Soon computer buyers, their financiers and project leaders of software projects adopted that view. They worked quite well for decades with this roughly estimated scale. |

| | | |
|---|---|---|
| 1970 | Lyndan B. Johnson, Spiros Simitis | On July 4, 1966 the Freedom of Information Act (FOIA) was signed into law by U.S. President Lyndon B. Johnson. Spiros Simitis, a German professor of Greek origin, was the first to put a privacy law into words. It became state law of Hessen (Germany) in 1970. Later the Federal Constitutional Court of Germany held it most important for citizens to know what data the different private and government organizations store directly or indirectly associated to a person. - At the beginning in situations in life, citizens voluntarily accept the registration of their data through social media or smartphones. Citizens may later discover with surprise the unprecedented intensity of the processing of their personal data. - The computer community discusses privacy, security and transparency with passion till today. A special chapter about privacy was integrated in the book. |
| 1973 | Robert E. Kahn, Vinton Cerf | When Robert Kahn and Vinton Cerf worked for the Advanced Research Projects Agency Network (ARPANET) they understood the abstraction that this early packet switching network opened. Their stepwise approach to a flexible and efficient networking technology changed the understanding of network communication. All proprietary networking concepts like e.g. the System Network Architecture (SNA) by IBM, the DECnet of DEC, the TRANSDATA Networking technology of Siemens and the concept of Open Systems Interconnect (OSI), published since 1983, became history as TCP/IP became as an open, smart and lean protocol a worldwide standard. |
| 1976 | Bill Gates, Paul Allen, Steve Ballmer | 1976 was the first year the company founded by Bill Gates and others had its official name "Microsoft Corporation". The products of Microsoft are probably daily used by most readers. Microsoft dedicates much information on the history of the company on its web site, so this space is used for other subjects (see: http://channel9.msdn.com/series/history). - Microsoft brought the PC to people's homes. The way this happened, possibly unintentionally, was later put into words by another computer company: Start small and grow fast. |
| 1976 | Steve Jobs | Steve Jobs had an extraordinary feeling for the user friendliness a computer needs. Instead of the "start" button of Windows he created computing devices that we could switch on and off like a lamp. The system functions were limited. Task handling, parallelism of the programs, and the interfaces were reduced. This simplified the use of the computer and thus made it |

| | | |
|---|---|---|
| | | attractive. The integrated enrichment with audio, video, shopping, mapping and many more functions made Apple one of the most expensive companies of the world. |
| 1977 | Larry Ellison | Larry Ellison co-founded Oracle Corporation in 1977 under the name Software Development Laboratories which, after more changes in branding, became Oracle Corporation in 1995. Larry Ellison understood the enormous chances of the relational database technology that E.F. Codd had laid out in his fundamental papers in 1970 and 1990.[27] The Oracle products soon became the standard of the industry and for the computer buyers. Open source database products, like mySQL, did not change this prominent market position. The Oracle database products were early enhanced for object oriented application programming by securing a strong compatibility. |
| 1980 | Charles Antony Richard Hoare | Today's programmers work in an industrialized, standardized and closed environment that secures regular office hours and salaries, lawful programs and projects delivered on budget and on time. They probably hardly image the strenuous struggles to arrive at this point. One of the pioneers, Professor Hoare, was a key contributor. ACM honoured C.A.R. Hoare in 1980 with its prestigious ACM Turing Award: "Professor Hoare was selected by the General Technical Achievement Award Committee for his fundamental contributions to the definition and design of programming languages. His work is characterized by an unusual combination of insight, originality, elegance, and impact."[28] |
| 1989 | Sir Timothy John Berners-Lee | Anyone who intends to document unstructured ideas finds himself in a fix, beginning at one point and then finding out that another idea is more fruitful. No matter what paper is used, this paper is not large enough to write down all ideas. Mind Mapping may be one solution. What is unquestionably needed is an open networked structure. The Internet with the http protocol is one solution, connecting world's knowledge. T.J. Berners-Lee managed to present the first communication via a client and a server computer with a basic Hypertext Transfer Protocol (http) in 1989. |
| 1991 | Ken Thompson, Linus Torvalds | Ken Thompson started at Bell Laboratories (a company of AT&T and Western Electric) the development of a new operating system. His intentions were to allow the concurrent processing of batch programs and interactive dialogs. 1983 brought UNIX System V to the market with training, |

| | | |
|---|---|---|
| | | documentation and professional support.[29] In 1983, several European Manufacturers cooperated in the joint improvement of UNIX, founded the X/OPEN Group that later merged with the Open Software Foundation to the 'Open Group'. Why did companies do that? In the following chapters the growing complexity of operating systems is explained. So the advantage for the companies involved in this cooperation is a tremendous decrease of costs and gain of independence (from products like Microsoft Windows). Cell phones and computers for a variety of machines need operating systems. UNIX versions are perfectly suited for that purpose. Why was the Open group not successful? Because they forgot to compete. - Linus Torvalds had another idea in mind: He wanted open software without commercial use. This limited the use of his work, as copyright legislation provides legal security only if a licence is given. So the General Public License (GPL) was introduced. GPL is now often used for open software products. |
| 1998 | Sergey Brin, Larry Page | The idea was visionary and fearless from a technical and commercial point of view: To provide for everyone's convenience a search index on all data published in the whole world for free. BusinessInsider published on April 24, 2014 "The Untold Story" of Larry Page, his management methods and Google. "*One day in July 2001, Larry Page decided to fire Google's project managers. All of them. It was just five years since Page, then a 22-year-old graduate student at Stanford, was struck in the middle of the night with a vision. In it, he somehow managed to download the entire Web and by examining the links between the pages he saw the world's information in an entirely new way. What Page wrote down that night became the basis for an algorithm. He called it PageRank and used it to power a new Web search engine …*"[30]. Page explained his reasons for firing project managers according to the same publication: "*… he didn't like having non-engineers supervising engineers. …* ". In the end, the project managers were kept and Google later hired more. In April 2014, the media reported that 1000 engineers were taken away from the Google+ social network project, as Google+ could not compete with Facebook.<br>According to the press reports, Google tried several management methods. E.g., in 2004 Larry Page and co-founder Sergey Brin started with a 1 $ salary. In April 2013 Google informed of the prolongation of this tradition for 2012. Thus they continued the practice that Lee Iacocca began in 1978 when he became CEO of Chrysler and Steve Jobs copied in 1997, when he returned in 1996 to Apple.[31] Page eventually wrote down his management |

principles. According to the above mentioned article in BusinessInsider, Page's rules were: (1) Don't delegate. (2) Interfere only if you add value. (3) Don't be a bureaucrat. (4) Juniors deserve the same respect and cooperation like seniors. (5) Don't stop someone with a simple "No.".

# 3. FROM IDEAS TO ARCHITECTURES OF COMPUTER SCIENCE

Monographs about the history of computing often date back to ancient Egypt and Greece. This is where understanding of Mathematics comes from. Mathematicians at all times had to teach others with simple recipes how to calculate, e.g. a bridge, a ship or a building. This is what we consider granted in this book.

Studying the history of computer science teaches a lot about the development of the ICT industry. Although we know that those living in a time of rapid development don't see the migration paths of old ways of living towards new ways of living clearly, it is obvious that only in the U.S. after World War II a systematic and combined effort to use the potential of ICT was taken. Actors of this effort were the universities, financial potentials, some politicians and of course fearless pioneers in all of these areas. Compared with the European strategies e.g. of the research and development program of the European Commission, the U.S. way to innovate is more efficient.

**Timeline Office Automation**

Christopher Lathan Sholes designed what "eventually became the first commercially viable typewriter in the 1870s. William Burroughs (1855-98) constructed an adding machine in the same period ...". [see: James W. Cortada, page 126] In the early 1950s, IBM delivered the first electronic typewriters. After the 1960s, the typewriters had a small memory so the last few characters, typed erroneously, could be corrected without spoiling the printed document. In the 1970s, scanned documents and other large binary objects were used to reduce paper archives and speed up administrations. In the 1980s, IBM released Distributed Office Support System (DISSOS) with Office Automation standards Document Content Architecture (DCA) and Document Interchange Architecture (DIA) with final form (like today PDF/A) and revisable form documents. DEC followed with All-In-1 as an alphanumerical office system, later with DECwrite on a What-You-See-Is-What-You-Get (WYSIWYG) basis. IBM's OS/2 was stopped when Microsoft won the market with Microsoft Office. In 2013, Office became available on Microsoft Phone and optionally synchronized through SkyDrive into the Microsoft Cloud (Office 365 or 2013).

Examples are found in the following text.

The history of IBM Germany goes back till 1910 with its title “Kleine Chronik der IBM Deutschland 1910 – 1987”.[32] James W. Cortada goes back in his book till 1930. Many foundations of Computer Science were laid before 1980. The most remarkable Research, Development and Application of the Information and Communication Technology started around 1980. The years before could be described roughly in two episodes: 1910 – 1960 and 1960 – 1970.

**1910 – 1960**

Looking back from today’s state of computing we view these years as a time of ‘search for digital computing’. Many ideas were born like e.g. the fundamental mathematical principles developed by Turing, but the fruits were harvested later. This was the time of fixed programs through plug-boards, of tabulation, and calculation based on simple arithmetic operations. Calculation was reduced to arithmetic formulas, e.g. through calculation of integrals using excessively long chains of additions with program loops. The electrical and mechanical parts, the tubes and the input/output operations were slow and not reliable. Computing reliability and speed are important, e.g. for military, for machine control, flying in space and later to the moon.

**Timeline Clinic Automation**

Clinic automation illustrates how computing changed professional work and financial management. First, medical diagnosis was documented on – mostly scattered – personal cards in huge drawers. Computing began with laboratory equipment; the diagnosis was recorded with patient’s names and birthdates to improve quality and avoid confused identities. Patient registration and billing was on a separate ICT system. Technology was first batch, then online read-only and in the 1990s, when transactions were supported, online-update.
1990 - 2000 systems integration began: Patient registration, electronic data of laboratories and each physician’s diagnosis was connected with the patients’ master data. Financial billing and analysis as well as archiving of medical and financial data were integrated after 2000. This became a basis for the definition of standardized quality assured treatment. Now all clinic personnel give patients optimal and efficient personalized treatment including all aspects of the patient’s disease. The use of HANA technology is the next step (see chapter 5).

The first versions of programming languages - FORTRAN for mathematical computing and COBOL for commercial computing -

were designed in the 1950s. Even the idea of artificial intelligence and first approaches through a programming language, LISP, was born in the 1950s.[33]

These times were years of fast technological development laying the foundation of electronics. James W. Cortada reports: "One important observation made repeatedly in this book is that the computer business remained a relatively minor one until the late 1950s and even then proved less significant than has been implied by other writers."[34]

**1960 – 1970**

The rise of computer science began: In 1962, Stanford and Purdue University created the first computer science department. Two years later, Thomas Kurtz and John Kemeny developed the simple Computer language BASIC. The computer mouse was invented the same year by Douglas Engelbart. In 1965, John Alan Robinson develops unification for logic programming which is one basis for today's automatic proof of mathematical logical hypotheses. Two years later the basics for object oriented programming (Simula) were defined by Ole-Jahn Dahl and Kristen Nygaard.[35]

Before 1960, IBM had (with its system 600 and 700) brought an understanding to large enterprises and the government that computing is saving cost and time. A new generation of engineers was educated and hired to develop the possibilities of the new technology beyond tabulation and processing of punch cards.

Germany introduced in 1959 the electronic bank account. General electronic money transfers started, e.g. for paying salaries to workers and employees, and for everyone to pay bills.[36] Without great computing powers, a standardized infrastructure for banking and money transfers would have been impossible.

The United States sent the first man to the moon. What a fantastic success, visible for everyone. Scientists believed in the universal possibilities of technology and well organized human endeavours! For more see the explanation to Joseph Weizenbaum.

In 1964, IBM announced its System /360. In 1968, IBM announces the system /360 Model 85 with the first cache memory, and in the summer of 1970 its

System /370. Other companies followed closely and were ahead in some respects.[37] Those were the decisive steps towards the general purpose computer. Companies like General Motors Corporation, General Electric Corporation, Siemens and Fujitsu now saw the chances and thought, they could estimate the challenges.

The organization of data and processing in an operating system was advanced with the concept of transaction processing. This will be discussed later with databases. IBM presented its first release of the Information Management System (IMS) in 1966 as a secure and restarting database. The networking add-on followed in 1969 with IBM's Customer Information and Control System (CICS). These systems were not easy to understand and to set up. Trained engineers were not available. So the competitors could follow later with e.g. UNIVAC's Transaction Interface Package (TIP) in the 1970s, Siemens with its Universal Monitor for Transaction Systems (UMTS) in the early 1980s and Digital Equipment Corporation (DEC) with its Application Control and Management System (ACMS) in 1985. These transaction oriented systems were first installed with index sequential file systems that possessed enhancements for transactions. Later, transaction enabled index file systems with several indices, variable data record length and flying restart capabilities like e.g. VSAM, IMS of IBM were given to the customers. The results were not sustainably satisfactory. More and more functions were required so the CODASYL and relational database concepts were standardized. Spin off companies, e.g. Software AG (founded in 1969) with former employees of IBM, became successful. Software AG developed a flexible high-performance database software ADABAS combined with the high level programming tool NATURAL. Both still hold a notable market share in 2013.

## 1970 – 1980

Integrated and interconnected computing became possible with new products like optical mark readers and magnetic card readers.

Data Security became an issue. IBM reports in its Internet Archive: "A two-year IBM-funded data security study, done in cooperation with MIT, TRW Systems, and the State of Illinois, results in data security guidelines for users."[38]

In 1971 the Canada Geographic Information System (CGIS) was released. Geographic data processing became one of the major sources of innovation for the information and knowledge society. The introduction of GPS, modern meteorology, digital planning of cities, buildings etc. became possible.

Geographic Information Systems (GIS) use different layers of information. The bottom layer is the geographical digital map. On top of this layer, content is registered (like e.g. the position of hospital, the borders of states or biotopes) using a suitable coordinate system. N layers are possible and used.

SAP was founded in 1972. Highly integrated software solutions of independent vendors or of departments within the big computer manufacturers were planned and released step by step. The idea to have "software out of the box" with all necessary functions was unrealistic: The software engineering model was to start small with one financially potent company, adapt the software package to some more customers and then make it and sell it as an industry standard. A lot had to be done: The structure of data records for data exchange (by exchanging media or by electronic data transmission through leased lines) had to be defined; processing cycles of producers and suppliers had to be synchronized. Delivery just-in-time was a promising idea coming up.

IBM announces Systems Network Architecture (SNA) in 1974.[39] Competitors follow some years later. The networking architecture depended on the manufacturer of the active networking computers. SNA with the IBM protocols (e.g. Synchronous Data Link Control (SDLC)) set an industry standard in commercial computing. DECnet, the networking technology of DEC, became a standard in many scientific areas and in universities.

On this basis, banks, industry, commerce, governments (e.g. for police and military applications) found a vision to develop online-update solutions. Corporate centers could now be informed "real-time" on a weekly or monthly basis about major business and administrational developments.

1988 the first fiber optic cable connects the U.S. and Europe via the Atlantic Ocean.

Barcodes were invented in the mid 1930s. Barcode technology provided early an opportunity to identify products and objects of any kind (like blood sample, patient identification in a hospital, book identification in a library) with optical recognition and reading of the code. In 1973 in the U.S. a universal product code (UPC) was used. Today's Global Trade Item Number (GTIN) is since 2009 an improved and globally used version of the UPC and the European Article Number (EAN). Today industry tries to substitute the barcode by radio-frequency identification (RFID). RFID allows to store and transmit more data and with a defined distance between the reader and the object through induction of electromagnetic fields. It is used today e.g. for ski pass, production lines, theft recognition and for paying with smart phones to substitute cash payment.

**1980 – 1990**

"The emergence of personal computing and networks in the 1980s set the stage for dramatic advances in computer graphics, software technology, and parallelism."[40]

Computing left the former world of alphanumerical and structured data. Geographical data, bit maps (binary large objects, digital pictures), documents and interlinked data could be processed and met high demand on the market. Database technology had to be enhanced to meet new requirements. Customers needed more speed and better performance, more table space, larger segments of data storage, inter-connected data bases, loosely and tightly coupled systems and failover with 7/24 easy administration.

1987 the Global Positioning System (GPS), developed in the 1970s, is launched and financed by the U.S. Department of Defense. If the GPS receiver catches the signals of four or more satellites, the position of the receiver is calculated. The current version of GPS became the world wide basis of modern navigation systems. The European Galileo positioning system, decided in 2003, is expected to be fully operational in 2019.

Innovation that once came from hardware from now on comes in high degree from software solutions. Software became the main competitive factor of ICT. The "IBM Systems Handbook" of July 1986[41] lists hundreds of software solutions e.g. for application development, database functionality, PC support, distributed systems support, plant management, document processing, graphical data processing, retail and supermarket and much more.

IBM developed in the mid 1960s an Airline Control Program (ACP). ACP was an application oriented operating system. Unlike previous such systems, the most notable aspect of ACP and the /360 computer family is that ACP was designed to run on most models of this mainframe computer family.

The computer industry became an indispensable industrial sector. Batch processing of cards was substituted by online processing. So registers like the police registers were built.

In 1981, the director of the German federal police agency ("Bundeskriminalamt"), Horst Herold, retired. He built up the ICT of the federal police in Germany and implemented the systematic search for criminals by combining two or more data bases and conducting a pattern search on the connected files ("Rasterfahndung"). Today the monitoring of criminals or terrorists is done through big data analysis and tagging according to what the law allows. The system automatically connects data records depending on various attributes (like face, handwriting) stored in the data records.

In the 1980s, the first efforts to standardize ICT service management framework were made. The reason was that the fusion of data centers and the contracting became highly complex as all data centers used a language and methodology of their own. A comparison of services was hardly possible, and standards of service level were re-invented by all service management teams. Till 1996 a library of standards was published under the name Information Technology Infrastructure Library (ITIL). This proved to be of great help to further develop data centres to profit centres.

## 1990 – 2000

The European directive 1999/93/EC of the European Parliament and of the Council of 13 December 1999 on a Community framework for electronic signatures[42] was published to create an e-commerce market within the single European Market and to help European companies. This objective was not met, as the European Commission was unable to harmonize the different strategies of the European Member States. Austria developed its 'Bürgerkarte" (citizen card, detailed technical information available since 2004[43]) on the basis of a health card. Germany developed its "neuer Personalausweis" (new identity card, distribution started in 2010) in another legal framework and at a time when other member states, like Spain, were much more advanced with their concepts.

In 1993, the idea to transmit voice with digital numbers instead of analog electronic waves was prepared for commercial use by standardization of the ITU. "ITU-T Recommendation H.323 was prepared by ITU-T Study Group 15 (1993-1996) and was approved under the WTSC Resolution No. 1 procedure on the 8th of November 1996."[44] The transmission of audio signals via the Internet, e.g. with the Skype service, followed later when the Voice over IP standards (VoIP) were defined.

September 4, 1998, Google was founded. Many people thought, the idea to index all servers and web sites of the world is unrealistic. Google showed that this was possible. It improved its services with Google Maps and Google Pictures. Details are offered by the company in the Internet. Other companies developed search engines. To be truly successful in this area, a lot has to be learned about "searching" in the electronic data bases. Search engines provide bad service levels without intensive customizations. The definition of the sequence of the listed search results, the synonyms and a lot of unexpected special cases turn nearly all implementations of a search engine into a difficult project.

1993, the Portable Document Format (PDF) was published. PDF/A is used for archiving electronic documents for future generations. Archivists hope that this format will sustain independently of technical progress.

Only the best software products survive the challenge of the millennium bug (Year 2 k (Y2k)).

## 2000 – 2013

A new era dawns with software products supplying rich functionality. Before 2000, clients waited for functionality urgently needed. The software vendors now offer software packages with functions in abundance.

The Internet became a reliable basis of international commerce and communication. Broadband access is a political issue. E.g. in Australia politicians decided to build a nationwide broadband network to make administration and health support easier. In Europe broadband is considered an essential infrastructure to motivate companies to establish affiliates in rural areas. Hotels rated "3 stars" need fast Internet access for their clients.

"Neutrality of the Internet" becomes a political issue in Europa after the Telecom companies developed strategies to differentiate broadband access costs by the real use of Internet traffic.

In 2004, Facebook was founded. Details are offered by the company in the Internet. Other social networks like XING for career support were founded, but could not meet similar success (status: 2013). Obviously Facebook invests much thought and energy to be the leading company. Users of social networks expect functions like building isolated groups of friends, being invited automatically to get introduced to new friends, being reminded about birthdays or new information from friends, automatic face recognition if e.g. the picture of a group of persons (like class mates) is published and people on the picture are users of the same social network. There are different opinions on the functions of social networks. As an example, some people don't want the automatic face recognition, and some people expect more privacy and copyright issues considered.

The EU services directive was published in December 2006. The member states were for the first time bound by law to implement all administrative services covered by the directive in classical and electronic ways. Since December 2009, all service providers in the European Union can receive specific authorizations when founding or changing a company by electronic means. An Internal Market Information System (IMI) was implemented to provide easy and automatically translated communication of authorization data of service providers between all relevant agencies within Europe.

In Europe, the public administration digitized the territory in several scales with advanced Geographic Information Systems (GIS). "In Europe a major recent development has been the entering in force of the INSPIRE Directive in May 2007, establishing an infrastructure for spatial information in Europe to support Community environmental policies, and policies or activities, which may have an impact on the environment."[45] The metadata definition used by the realization of the INSPIRE data bases is used as a basis for Open Data solutions after 2010.

.

Those who want to extrapolate the history of computing science may find a mirror of what is coming in the ACM[46] Tech New of June 7, 2013. It reports the data mining activities of the National Security Agency, the fact that WiFi signals may be interpreted through the Doppler frequency shift in order to allow the control of devices through gestures and the thought-powered helicopter flights through a field of obstacles with 90% success.

## TODAY: HIGHLY INTEGRATED SOFTWARE SYSTEMS

2005 to 2014 big multi-functional software systems became of a quality and reliability which was never expected to be possible. Such quality was necessary in order to develop cloud computing and to maintain the software for the support of large customer bases e.g. for content stores or search engines. Although software providers like SAP, Microsoft made enormous efforts to unify and stabilize their software, especially to meet the requirements of large clients from industry, the competition arising from the cloud technology brought new levels of quality to the ICT market.

Cloud computing arose as a follow up technology from shared computing services, server vitualization and computing on demand with three functionalities: a) the billing of computing was according to the computing power used (with several different billing models), b) the cloud provided its users with potentially unlimited computing power and c) the user of the cloud could customize the software on its own without help from the software provider. IBM was one of the technology leaders in cloud computing. Navision, a provider of accounting software with further functionality, was one of the leading providers of software which encouraged their customers to customize the software on their own. (see the chapter on cloud computing for further details).

Driven by the expectations of end users and share owners, the providers of integrated portals and general services increased the functionality of their services. A short and incomplete description shows what was achieved (mostly in the years 2000 till 2014) in the area of globally used applications:

- Apple is known as the ultimate provider of user friendly personal computers, smartphones and other mobile devices. The company offers not only such devices but showed a substantial groth on the basis of content provision (music, video, etc)..

- Google operates a world-wide search engine for among other objects text, pictures, books, news, mailing and office solutions and social information. It integrates these modules with the Android operating system for smartphones and tablet computers. Users of such mobile devices get a cloud services for the upload of digital fotos and the synchronization of mail, calendar and further data.Google provides a picture processing software Picasa for a free download to search and modify digital pictures.
- Microsoft started in 2014 to distribute its operating system for smartphones as free software. Delivering Microsoft office for iPAD of Apple Inc. is another noteworthy change in the history of the company that for long was the only provider of PC operating systems and office software. For companies Microsoft provides software for operating systems, cloud applications (like Azure), web presentations, groupware, open government etc. PC users get free download e.g. for security modules and other helpful software.
- Amazon, Ebay, Facebook and others developed their software rapidly and offered not only end users a huge functionality but also commercial customers a virtual market place, where they could start an online shop. .
- Shutterstock, Pixabay and others set up portals for digital pictures either on an open community concept (like Pixabay) or as a commercial service.

A complete or a more detailed history of these highly integrated application services with illustrations of the functionalities would be very interesting but it is currently not allowed by the law (e.g. copyright law). The reader should refer to the history presented by the companies. It is mostly in the Internet.

## ICT ARCHITECTURE

To understand the coherent research and engineering development in different areas and the progress achieved we begin looking at the mainstream ideas that we call nowadays "architectures" in computer science.

But many ask: Why do we talk about architectures? Doesn't that sound a bit arrogant? The answer is that the word 'architecture' describes well how one idea is built on another. Of course, this is what mankind does for centuries. Of course, this is why writing and documentation was invented and used. But, since the 1980s computing added new dimensions and gave this power to all of us. If

I hit a key on my keyboard in London, UK, my friend in Perth, Australia, will see it 'at the same time' in his home in Perth, Australia.

What was the initial birth of architecture? Leibniz described the calculator and invented the binary digits. Boole added binary logic. Turing created the Turing Machine with the formal and stepwise processing. Babbage had the idea of conditional programming statements, and John von Neumann put all things together later with the Integrated Circuit (IC) and the Microprocessor (µP). That was already a rather complex framework and allowed to build a powerful machine. - Readers who like to understand hardware from scratch should refer to the available monographs.[47]

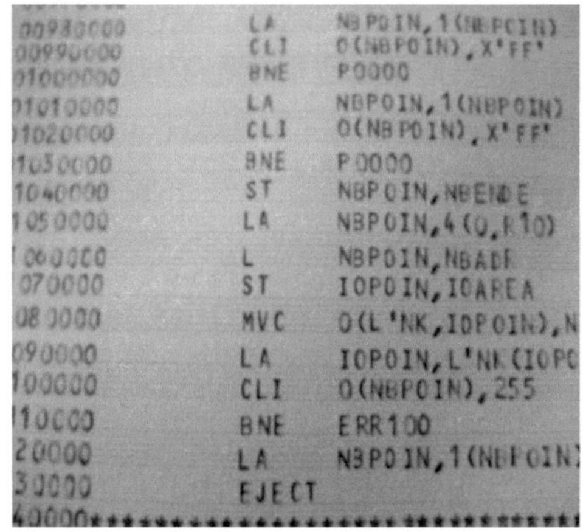

Fig. 4: Example of an Assembler Program

First, the processing unit needs to be explained. It consists of storage cells (they are called registers), especially fast, and enhanced with functionality like addition or multiplication for integer and floating point numbers. Even the early machines got several registers, R1 to Rn. The storage for the data and the program was small or large, at any rate the cells had to be numbered with addresses starting from 0. This gives – again in pseudo code – a program to add the first k numbers (k is an integer). The instructions of the program are part of the "machine language". The computer stores the instructions consecutively in the storage as well as the addresses, indicated here by A1 … Ai. The text after the "*" is a comment which is not seen by the machine. This kind of programming language is called Assembler Language.

A1: Read k from punch card to R1
A2: Read 1 in R2 * R2 counts the cycles
A3: Read 1 in R3 * R3 is the number 1, 2, 3, …, k-2, k-1, k
A4: Clear R4 * R4 is the result of the addition
A5: Add R3 to R4

A6: Add 1 to R2
A7: Add 1 to R3
A8: Compare R2 with R1
A9: If R2 is bigger than R1 (i.e. k), goto A11
A10: Goto A5
A11: Punch the result R4 to a punch card.

As a shorter way to describe arithmetic operations, we write

$$R3 := R3 + R4.$$

This means that we first calculate the sum of the contents of R3 and R4 and we then store that content in register R3. So, ":=" means "is stored in".

The architecture becomes more complex with data structures used in storage in order to process more complex problems of daily life. Architecture then becomes a theory: Computer Science.

To formulate a theory, mankind always needs a notation to document it. We here use the following system for data structure definitions: A data element D, which for simplicity sake we consider to be of constant length, is stored at an address which is denoted by (D). If a data element consists of several data parts, e.g. D1 and D6, we write (D1, D6) to denote that this is the composite data in the storage with the data length defined as the length of data D1 and the length of data D6. If we want to order data in an ordered list, we need a pointer from one data element to the next. This pointer is an address, which we denote by A1, A2, …. So we could say A1 = (D).

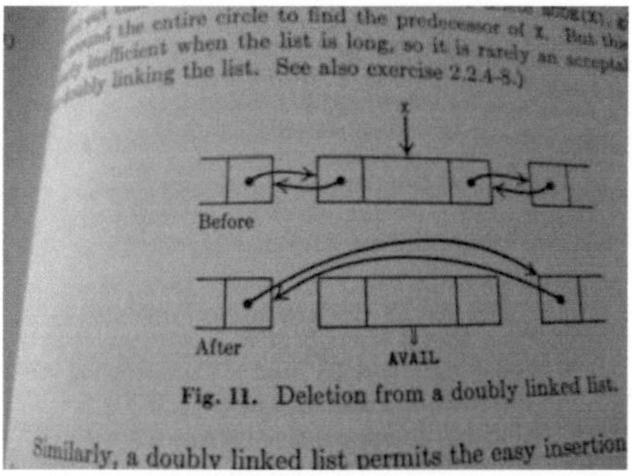


Fig. 5: Donald E. Knuth's explanation of a deletion from a doubly linked list

With this notation, a sorted list looks like that:

$$(D1, A2) \rightarrow (D2, A3) \rightarrow \ldots (Dn, A(n+1));$$

to indicate the end of the list we set A(n+1):=0

or

$$A1 = (D1, (D2, (D3, \ldots, (Dn, 0)))\ldots).$$

.

Now, the reader can see that the description of operations on these data is rather lengthy, although the concept itself is easy to understand using graphical notation. Graphical notation is more difficult to use for formal considerations like the performance of sorting algorithms or category of the operations needed. The term

A1 = (D1, (D2, (D3, …., (Dn, 0)))…)

is obviously recursive. This implies a specific program structure to deal with it. Recursive structures and processes are frequently found in problem solving in Computer Science.

The most important operations for structured lists are search, up-date (i.e. integrate new elements, delete elements within the list) and create a new list, sorted in another way. The reader is invited to invent some strategies. Sorting was in the 1960's an occupation that had its own experts (e.g. for QUICKSORT[48]). To feel the challenge suppose the data D1 … Dn are composed of data parts. Di could be name, address with postal code and / or street name of customers. So sorting is necessary for delivering goods in the fastest way.

Important and more complex was and still is the indexed list. Here, data is stored in semi-ordered blocks according to sorting criteria (e.g. name) which is part of the data structure (e.g. in a file of data about salesmen of a company). The data blocks are also sorted by that criterion. To find a record in a file, that file is not read sequentially but by using a shortcut with indices. If the search criteria are between two indices, the program knows that the data is stored in the data blocks that these indices point to.

Using the notation above, we can illustrate it like that:

| 1st level index block | 2nd level index block | Data blocks (unsorted within criterion interval) |
|---|---|---|
| (highest criterion of all data blocks covered by index block 1 of 2nd level address-index block 1) | (highest criterion in data block 1, address data block 1) | D1, D2, D5, D6, D3, D4 |
| | (highest criterion in data block 2, address data block 2) | D7, D8, D12, D10, D11, D9 |
| | …. | |
| (highest criterion of all data blocks covered by index | (highest criterion in data block n, address data block | D…. |

| | | |
|---|---|---|
| block 2 of 2nd level address-index block 1) | n) | |
| | (highest criterion in data block n+1, address data block n+1) | D … |
| …. | | D …. |

Such data structures were needed, when the data of thousands of customers had to be checked, changed, up-dated. The first commercial applications were applications with personal data, e.g. payments of salaries, customer files, financial files of banks, tax registers, citizens and car registration.

Donald E. Knuth, Stanford University, wrote several volumes about "The Art of Computer Programming"[49]. He was one of the experts, who also underestimated computer science document all aspects of programming. The reader finds in chapter 2 of volume 1 "Fundamental Algorithms" the description of the data structures. Although D.E. Knuth needs more than 200 pages to describe the information structures, known and needed at the time, most readers might agree that the basic concept is rather simple. Here is also much room for further inventions.

# 4. THE ERA OF GENERAL PURPOSE COMPUTERS

The explanations given to the achievements of the early pioneers of ICT and microprocessor development in Chapter 2 make us understand why there were so many doubts about the best technology. When the 4, 8 and 12 bit microprocessors were tested, when military, financial institutes and government could clarify their needs and when hardware performance and software concepts converged, the time for the general purpose computer came. Several companies working in the field of analogue devices or electricity now discovered the chances of the digital hardware modules and their integration.

**IBM and the seven dwarfs:** The 'dwarfs' with about 30% market share were General Electric (GE), Radio Corporation of America (RCA), Burroughs, National Cash Register (NCR), UNIVAC, Contral Data Corporation and Honeywell Information Systems (HIS). In the mid 1970's GE and RCA sold their operations, so the 'BUNCH' was left. Fujitsu, Siemens, Amdahl (1970), Comparex, Compagnie des Machines Bull (Bull), Data General joined in and sold computer hardware, software and support. At the time operating systems and standard programs of manufacturers were functionally poor, so their customers paid for the development of solutions. The computer companies tried to generalize custom specific solutions and incorporate them into their portfolio. Project management and quality were often poor. Only big companies like the computer manufacturers could train their employees to become outstanding computer managers.

'Exotic' companies like Intelligent Database Machines (IDM) and Telefunken produced innovations but could not grow sustainably.

In 1972 SAP was founded in Germany and Cray Research in Minneapolis. Oracle was founded in 1977 inspired by the publication of an article of Edgar F. Codd in 1970 on relational databases.

The early 1970's was the time of the 16-bit microprocessor. It was the time of PDP-11, Honeywell Information Systems H 316 and IBM's /360 product line that slowly emerged. The market was highly competitive among "Snow White", which was IBM, and the seven dwarfs (see box). The companies struggled with technology, project management problems and high competition that dropped the cost of computer hardware in the 1970s dramatically.[50]

Their basic architecture became a state of the art. To draw a picture of these times of highly motivated and fast innovations, we consider the 1970s and the early 1980s.

The ***central processing unit (CPU)*** possessed several fast memory locations that allowed digital operations like addition, multiplication of integers and floating point operations, instructions to move main memory areas, conditional and unconditional checks and jumps within the program. The CPU controlled the input/output devices with several interrupt handlers: Whenever a device reported a change of the state of operation (like e.g. from 'waiting' to 'busy', or from 'ok' to 'error', or 'division by zero') the CPU was stopped (interrupted) at the current instruction and a special interrupt handler program started at a specific memory address in order to secure the overall functioning of the system. A semaphore instance was installed. It represented modules, that only one program could work with at a time. Each semaphore was (and still is) used

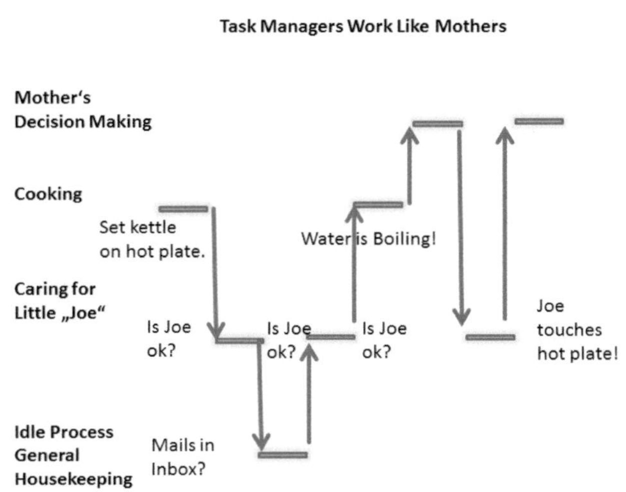


Fig. 6: Principles of Operation of Task Managers

by only one program, no matter how many interrupts were pending. Synchronous and asynchronous events and messages are reported from one task to another, if necessary.

A ***main memory board*** was installed. On specific low address locations resided the system specific modules, like drivers and interrupt managers. The upper portion of the memory was used for one or more application programs.

As soon as the memory was big enough to store several programs in it, a parallel execution of programs was intended. For this purpose, the n application programs and the system programs had to be controlled by a ***task handler, task manager, task scheduler or hypervisor***. The notion "hypervisor" was often used for task handlers managing different operating systems on one single machine. Task Manager's software modules still exists in today's computer systems and coordinate all parallel computing actions. It is quite a complicated piece of software, as strategies are complex to work with several CPU's, to run hundreds or thousands of programs according to their relative priority, and to deal with a variety of overload situations. It became even more complex when telephony, computing and networking were integrated in one system, that probably had to consider resources on other computers in the network as well. – The programs thirst for more and more memory grew exponentially. Sometimes, e.g. when the general purpose computer was used for data entry, the "user help"-module requested more memory space than the productive processing modules, but those modules were rarely needed and if, it hardly mattered how fast the help text came up. So the idea came up to build a virtual memory[51]: first as overlay modules, that could be called as a subroutine from the application program, and later as an integrated part of the operating system. When the program was started, it was dynamically linked and loaded into the upper part of the main memory. Main memory was segmented into segments of fixed length, e.g. 2k bytes. Each segment had a timer associated with it.

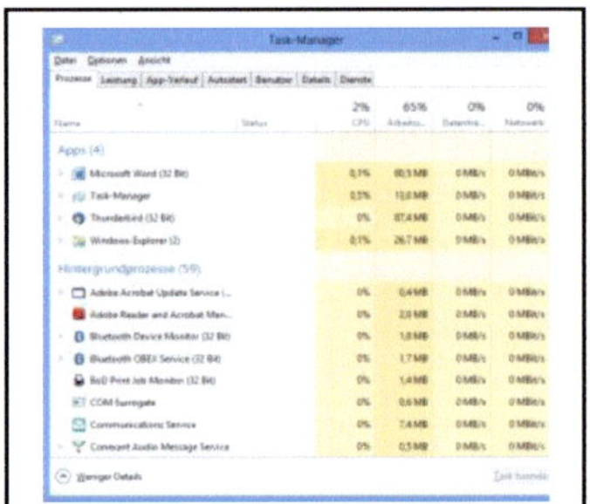

Fig. 7: Modern Computer Systems need also Task Managers (TM). Here: TM of Windows 8

When an application program jumped into a segment of a program (by "goto" or "jump") that was not actually loaded in the main memory, the following process started: The segment addressed by the "jump" instruction was loaded. To find a place where to load it to, the virtual memory management looked for the segment of the main memory that was unused for the longest time.

The task scheduler is different from the ***job scheduler and the printer scheduler***. The job scheduler is given to the organizers of the computer operations so they can plan and run financial programs, salary and sales calculations etc. according to the needs of the departments. The printer scheduler makes sure, that (i) pages can be reprinted if torn or badly printed, that (ii) only the required number of copies and pages are printed, no matter what happened to the printing process, that (iii) only authorized personal gets the printouts at specific places and printers and especially that (iv) the legal requirements of the financial market and of taxation are met.

Input / output devices of all make and speed must be controlled by the CPU and the application and system programs. Each device may have another method to control it, groups may be put together, so ***specific device control programs*** ('device drivers') are installed and called by the system programs as a module or as a kind of subroutine. The different schedulers need to be informed of error situations. Error information is routed to the application that caused the error. If real-time help is not available, restart of interdependent job sequences is managed by the job scheduler.

Last but not least, cables connect CPUs, memory, devices and the terminals of the operators or users. These cables are called ***system bus***. The number of wires of the bus determines the data transmission capacity. This bandwidth defines (i) the transmission mode of the bus (duplex or half-duplex), (ii) the speed of the bus, and (iii) its tolerance or intolerance through error detection and correction bits. Whenever information is transmitted via the bus, it may become erroneous on the way. These and many more characteristics make the system bus of the general purpose computer play a decisive role for the performance. Soon the bus defined the interfaces for the third party vendors, e.g. of fast disk vendors like EMC Corporation as well as for competitors that wanted to sell specialized devices only, like automatic teller machines. These Plug Compatible Manufacturers (PCM) became an annoyance for IBM.[52] During the struggles of IBM in the famous anti-trust case, that was opened on January 17, 1969, the position of the PCM became stronger as IBM stopped bundling its products.[53]

No doubt, for such interaction of highly complex modules 'system architecture' is needed. To handle data and information structures multiply indexed lists were needed. Soon the principle was stated, that redundant storage of data like "customer address" is always error prone. All data should be stored only once, kept secure and permanent accessible for all applications: The idea of the database was borne. Of course, these systems also had to be managed: The system management tools were added. Quality aspects like reliability gained importance. Hardware was improved, e.g. with Error Detection and Correction (EDAC) memory modules, Redundant Arrays of Independent Disks (RAID) storage were invented with different levels like RAID 1 to RAID 6. Restart techniques were incorporated in software, and control interfaces guaranteed proper access control.

Computing architecture is becoming more and more complex during the development of Computer Science till today. Back to the 1970's: The notion "operating system" (OS) was used. Honeywell built the General Comprehensive Operating System (GCOS) and MULTICS[54], IBM built OS 360, Virtual Memory (VM) and Multi Virtual Memory.

These complex systems needed large teams to plan, build and run. Thomas J. Watson reports that IBM underestimated the immense effort to build a general purpose computer and it's software when producing the /360 computer family.[55] Research always gave the basis for the later developments of the engineers, like e.g. with Edgar F. Codds paper on relational databases. A lot of scientific research was necessary to come to concepts that really worked in daily life.

One field of science is project management. To bring a new product into the market, in time and in the budget, was a problem from the beginning. New ideas had to be understood like the detailed and complete specification. The critical paths of the milestones of project management were found and people were trained to use the project management theory in a methodical way.

## CENTRAL PROCESSING UNIT (CPU)

The successful designers of the early central processing units, CPU's, were convinced that CPU speed, microprocessor cost and storage size would improve rapidly. Supposing a computer should read several sorted input data streams from n card readers, print data records, recovering from errors on a printer and store the merged data streams on a tape, the question is how to design this computer configuration? The first idea was to give the CPU full control. Whenever the CPU read data, the data path was via the CPU to the main memory and, after some calculations or other processing, back from main memory via the CPU to the output media. This way to control peripheral devices was called Direct Memory Control (DMC).

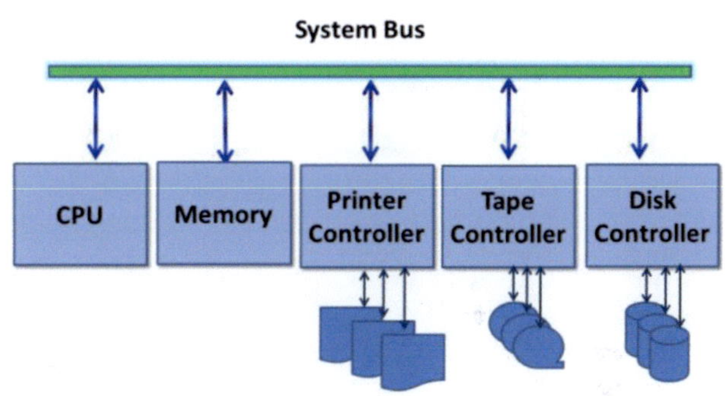

Fig. 8: Successful Model of the General Purpose Computer

Multiple problems for the early computer engineers existed: The input data and the data in the storage need a code for the binary data. For the microprocessor in the CPU and for the electronic circuitry in memory and device control elements, a register size had to be determined. Furthermore, an instruction set is needed: A simple instruction or a complex instruction set must be defined. Some designers did not have the courage to implement one instruction set for scientific and commercial problems.

The designers needed to know how programmers think and how they intend to use such instruction sets. Empirical studies were necessary.

Next problem: The input and output devices are controlled by the CPU. How? The idea of advanced designers was to design specific microprocessors as device specific controllers. These device controllers are connected to the CPU via a data bus working in duplex mode. That allows data and address transfer in parallel. The clock cycle for this bus determines the speed of the system. This technology was used with prudence as the costs and the many asynchronous events, that had to be handled, were not understood. A lot of theoretical calculations, practical simulation and module tests were necessary. Finally, within about 15 years, the peripheral device controllers were microprocessors that could handle data transfer on their own to and from main memory. This was called Direct Memory Access (DMA).

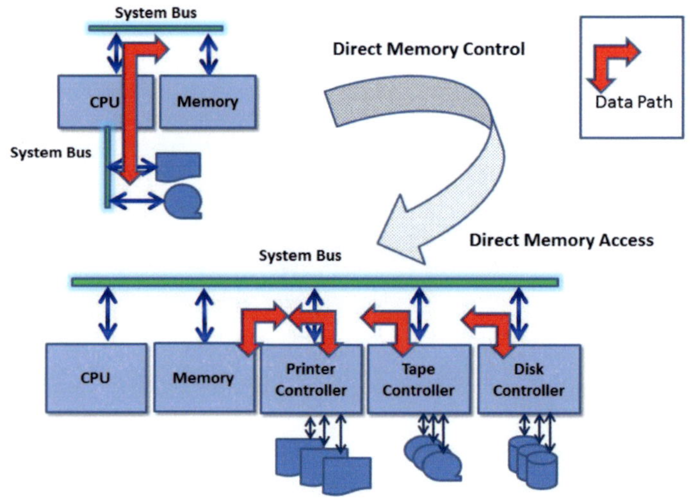


Fig. 9: The technical progress from Direct Memory Control to Direct Memory Access

The implications were enormous as asynchronous actions had to be handled. If, as an example, the CPU calculates a table with program A and in the same time the microprocessor or card input initiated by program B reports ('interrupts') that new data arrived in the main memory. The CPU can't ignore this. Consequently, the CPU stores the processing status and environment of program A. The CPU re-loads the status and environment it used when initiating the read operation to the card reader, and then the CPU resumes its work for program B. That may not look too difficult. As soon as there are hundreds of programs A1, ... A100, and as soon as during the re-load for program B there is another interrupt, the CPU needs more functions. First, it needs a privileged processing mode that cannot be interrupted. Second, it needs a data structure for fast un-loading and re-loading in main memory that is tightly connected with the hardware structure of the CPU.

Innovation was rapid. In 1964, IBM shipped its new /360 computers. /360 models were equipped with a 16 bit register set for the cheap model and a 32 register set for the other models of the mainframe family, coded with 8 bit EBCDIC per character, a minimum of 4 k Byte main memory (in 1971 up to 32k Bytes), and a complex instruction set designed for scientific and commercial problem solving. The publication "IBM System /360 Principles of Operation", which is available in the Digital Library of ACM, describes the architecture in greater detail.[56] The chief architect was Gene Amdahl, a courageous and innovative engineer and scientist.

In 1968, Intel ("Integrated Electronics") was founded by Robert Noyce, Gordon Moore and the visionary Andrew Grove. Their first microprocessor introduced in 1971 was the 4 bit µP INTEL 4004. In 1972, the INTEL 8008 was released as an 8 bit µP which could execute 50.000 instructions per second. The INTEL 8086 was the first 16 bit µP with a 16 bit data bus and a 20 bit address bus. It executed up to 2,5 million instructions per second. The INTEL 80386 was introduced in 1986 as Intel's first 32 bit µP. This innovative development continued and in 2000 the INTEL Pentium IV was released with a clock speed of up to 3.8 GHz. In 2006, the INTEL offered a 32 or 64 bit µP as the Pentium dual core processor.

Other companies developed comparable families of microprocessors and computer mainframe models. IBM developed an additional Reduced Instruction Set Computer. Honeywell Information Systems (HIS) with the HIS 316, HIS 516, HIS 716 and later the Level 6 Minicomputer, Digital Equipment Corporation (DEC)[57], Hewlett Packard Company (HP) developed their own processors and mainframe families.

Over the years, HP published in its Hewlett-Packard Journal the design and development concept and the resulting architectures. For readers with deeper interest, this is an easy to read professional publication. - In 1985 HP gave the development project for its new generation of computers the code name "Spectrum, because the new computer line will include a spectrum of models ranging from desktop workstations to mainframe-class machines, all based on the same architecture."[58] The architecture, later called "Precision Architecture", was based on the analysis of billions of instruction executions to determine the optimal set for the new computers. The designers used a RISC architecture with Very Large Scale Integration (VLSI) chips. In January 1986, HP introduced some compilers, in August 1986 the "HP Precision Simulator" and in December 1986 the HP-UX, the HP Unix Operating System with Data Base Management. In September 1987, HP Precision Architecture (PA) was released with VLSI Processor[59] and an N-MOS III Single Chip CPU. The HP PA provided up to $2^{32}$ distinct address spaces, each up to $2^{32}$ bytes long. The RISC processor, which usually aims to execute one instruction per clock cycle, was enriched with a floating point coprocessor.[60]

Today RISC processors drive mobile devices. They are integrated in mobile cell phones and smartphones. The architecture of the RISC processors began with the IBM 801 architecture and soon continued in the universities.[61] Today the ARM[62] processor is a prominent example for this technology.

Technological progress is breath-taking. DEC's Dick Sites and Fred Brooks started in 1989 to develop a new RISC processor called Alpha for the company that won its money with the CISC based VAX/VMS operating system. At the time, the fastest RISC machine was Intel's N10 with 40 MHz clock cycle. In parallel, IBM introduced its RS/6000 RISC computer family. The 64 bit chip architecture was intended to last 25 years and to allow a 1000-fold performance increase to 400 billion instructions per second. At the DECworld exposition in April 1992, DEC presented a prototype system based on a 150 MHz Alpha chip. At this time, HP announced its Precision Architecture 32 bit PA-4 with a peak of 66 MHz.[63] The Alpha chip was a success and allowed e.g. movie makers to start in new dimensions of virtual reality and animated movies. Alpha chips could not save the autonomy of DEC.

Instruction Set

The microprocessors were invented by many manufacturers on a proprietary basis, as no industry standard had emerged at the time. The von Neumann computing model implied the necessary instructions: One family of instructions was needed to shift Bytes (characters and numbers, integers and floating point) in the main memory from one location to another. Bytes had to be compared: A conditional jump (branch) instruction gave the programmer a possibility to act according to the result of the comparison. A further family of instruction was implemented to allow numerical calculations. Peripheral devices needed control, which happened through supervisor calls.

As now experience was at hand what instruction set was the best for applications to come, several designs had to be adapted to successful machines and to the competitors instruction sets. Firmware provided the solution. Especially Honeywell Bull emulated with firmware its machines from different origin (Compagnie des Machines Bull, France, and Honeywell Information Systems, USA).

Later several manufacturers released virtual memory machines that could emulate two or more different operating systems controlled by hypervisors. The effect was often a processor working a notable percentage of the time in an overload mode. Overload modes are poorly handled by task schedulers and hypervisors. Often low performance resulted from this architecture. As today CPUs are much faster, this effect is no longer a serious problem of the data centre.

## COMPUTER ARCHITECTURE AND ITS IMPLICATIONS

The development of the computer architecture went roughly in the direction outlined in Fig. 7. Details and advantages of the various manufacturers allowed access to different markets at the time of the introduction of their computer systems. As an example, DEC offered early a smart way to connect computer systems in networks as clusters.

The model of the general purpose computer made a lot of sense: The CPU working in microseconds is much faster than all peripheral devices with response times in milliseconds. The CPU processes a lot of calculations while waiting for the termination of input and output operations. The best overall throughput might be achieved with a mix of applications. At least one program computes a lot of calculations to ensure high CPU use, and several programs have dominant input and output activities while using little CPU capacity. Running different applications at the same time requires safe program isolation while sharing the main memory, devices and CPU between those programs. These requirements lead via overlay techniques to secure and isolated virtual memory in an operating system.

Once again, much more is needed. The CPU plays two roles: First, the CPU manages as supervisor the different programs and the input / output devices. Second, the CPU is a resource itself that the CPU as supervisor attaches to the different programs for temporary use. This temporary use is terminated by events, like the termination of an output operation of program A. Events are handled by a supervisor program. It is called 'Event Handler' and seen as a functional part of the Task Scheduler (Task Handler, Task Manager).

The Event Handler manages the interrupts initiated by the system watch dog clock and the asynchronous feedback from the device controllers. Why does the watch dog come in? Gödel said that we cannot create error free code. So, program code might run in a loop. Only a watch dog timer can get the CPU out of the loop and give that precious resource back to the computer system to continue its operations.

Another event that does not originate from peripheral devices is the division by zero. The necessity to cope with error situations is another reason for a privileged (non-interruptible) and a normal (interruptible) processing mode. This short reflection on the basic model of the general purpose computer shows the implication of this advanced model.

Over the decades, the disk controller was enlarged to a Storage Area Network (SAN) working with RAID technology for servers of a server farm. Supplying storage to several servers and mainframes, the SAN manages the external storage, its back-up and maintenance in an advanced full service solution. The single CPU got a second, third and even thousands of additional CPU's. The system bus became the core of high speed network architecture closely connected to server modules. Several further developments opened the way to cloud computing (see the chapter on networks and the chapter on cloud computing for more details).

Specialized manufacturers and pure software companies emerged. The manufacturers could not sustainably secure their originally occupied markets during further developments. New concepts and the cost advantage of the highly standardized processors allowed computer users to "downsize" to cheaper processors. Those standardized processors encouraged many software developers to produce highly productive software. Cheap processors were developed and sold in parallel with productive development tools. The results were often revolutionary new applications.

After 2010, the observers of the ICT market became convinced that further progress and revenue will come from software development in the future.

## OPERATING SYSTEM

The general purpose computer became complex during the 1970s and needed specially trained engineers or scientists. These people were available in limited numbers. Banks, government for revenue processing or citizen registration, industry for production control, airlines for ticket sales and flight schedules could not find people with sufficient training. So, two things happened. First, nearly everybody was hired and got training (e.g. teachers, engineers of any profession, even bakers) that showed interest in computers and was willing to work excessive hours in front of an often unreliable computing machine. Second, the computer was made less complicated by an operating system. The different modules of an operating system are now introduced.

### Main Memory

Main Memory of the computer is till today an expensive module of the computer. The most expensive module is the CPU. CPU and main memory define the performance of the system.

General purpose computing requires an architecture that satisfies the following conditions:
(i) A high number of programs should run in parallel with different priorities. As main memory can never be large enough to store all programs simultaneously in this memory, swapping (or paging) mechanisms are needed to bring pages from external fast media like disks to main memory and vice versa.
(ii) Fast and slow input/output should be supported. Especially communication lines became fast. So an enormous amount of data flows in main memory and needs buffer space. After processing, these buffers are free again. This management is called buffer management.
(iii) Zero breakdowns are necessary because of the many users and the critical applications. So, error detection and automatic 'flying' restart is needed.

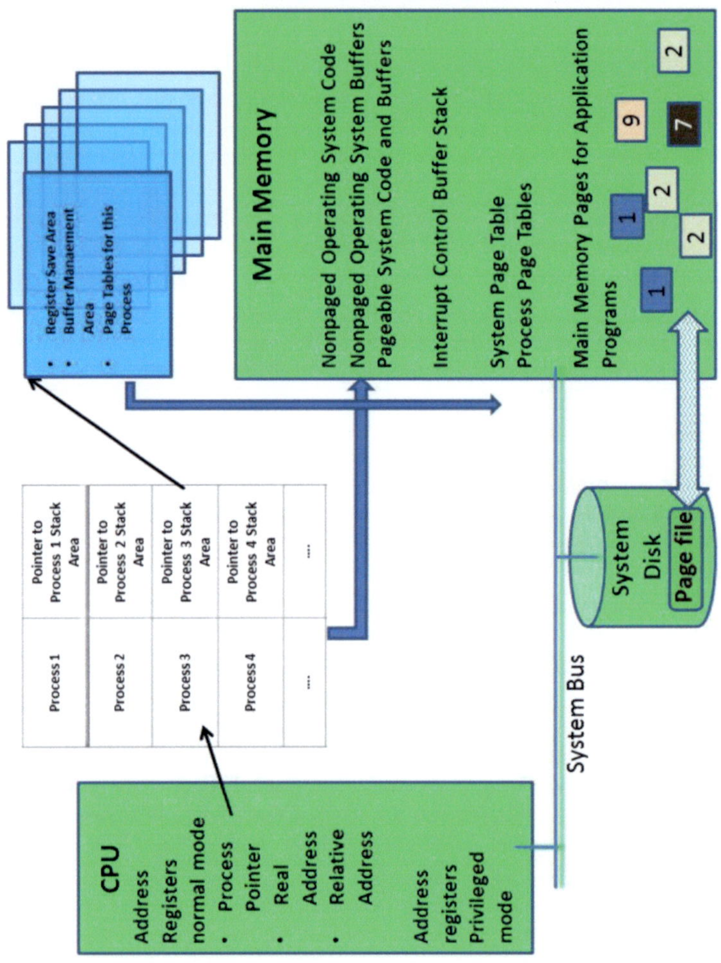


Fig. 10: Outline of Memory Management

Fig. 10 is a rough outline of the memory management. To explain it from a functional point of view, a process (or: task) is defined as the object in an operating system (OS). The identification of this object is used by the OS as an identity of a user application like e.g. the accounting program or a monthly salary calculation and credit transfer to the bank accounts of each employee. All application programs run as separate processes or as an identifiable group of interconnecting processes in the OS.

When the OS starts a process, a new environment for this application program is created: The program code is loaded in main memory. Input/output devices are assigned.

Device assignment was at the beginning the reservation of a real device like a card reader. Soon, spooling was invented. Output data were stored in a temporary file which was handed over to a system program of the OS, the spooling system. This spooling system directed the temporary files to the output devices, so the physical device was assigned to the application program only temporarily and with great flexibility. As an example, the invoices could be printed to a file of the printer spooler. Any time after the process was finished the real printout began under spooler control. Each process defined an application program, a set of devices, a program priority within the system and an isolated sharing of the memory.

Sharing memory is dynamic. For this purpose, a substantial part of the memory is reserved for application programs and split up into fix length segments. The length could be 2 k Bytes. When the application program is loaded in main memory, the loader of the OS looks for free segments (or: pages) in this part of main memory, allocates each segment (page) to the process memory table and leaves it to the dynamic memory management. As this loading and segment allocation is done in parallel to more memory space than is really available in the system, the memory management dynamically writes those segments (pages), that were not used by the processor or the device controllers for a "long" time, to the system paging file. This mechanism allows all application programs running in parallel with the illusion of a capacious main memory.

Furthermore, the allocation of each segment to one and only one process memory table isolates the virtual process memory of application program A from application programs B, C, D, E etc. The dynamic swapping of the oldest and not used segments of the whole main memory to the paging file guarantees the best overall throughput as far as memory is concerned.

As far as the CPU is concerned, the best throughput requires many event driven changes of the processes by the CPU. If e.g. a program A branches to a segment that is not in main memory but in the system paging file then the CPU initiates a read operation from the disk controller to get this page. The address reads the CPU from the process specific memory table. This may take about 100 milliseconds, a lot of time for the CPU. So the CPU unloads it registers to the process specific control area in system memory, looks in the list of eligible processes to the one with the highest priority, re-loads its registers from the process control block of this process and continues the calculation. CPU receives an interrupt signal when the read operation for the segment of the first process is successfully terminated. Now this process is eligible as a possible candidate for calculation, and the CPU may continue to calculate its program.

When virtual memory and the paging file were first released by manufacturers as part of the operating system functionality, customers valued it as an expensive feature. The reason was the overhead that paging caused for the CPU. Additionally, for the first applications (like data entry systems, real-time processing in production control or the processing of large batch programs) a lean version of paging, the overlay technique, promised less cost. Overlay management was a function of the operating system. It allowed the programmer to swap parts of memory with the contents of a disk file. The programmer could control all input / output to the overlay file and did not depend on the early (sometimes unreliable and not efficient) paging. Overlay technology was soon forgotten as paging became fast, comfortable and the costs were reduced according to Moore's law.

Memory management does not only manage the paging file and the virtual memory. Buffer management is another import work for it. Application programs should only in rare cases use static buffers in the program code as this reserves static buffer area, whether the buffer area is actually used by the program or not. Good programmers use dynamic buffers. If an application program needs a buffer area of 2 k Bytes, the program calls the buffer manager of the OS and the buffer is provided in the requested size through a pointer in a register. There is a further advantage: In case of an error, the buffer manager recognizes the buffer overflow operation of the faulty application program and the OS can terminate the program.

Re-entrant programs are programs that systematically do not use buffer areas defined and reserved within the program code. They always anchor the buffer

.

assigned in a process specific control block. These programs can be interrupted at any time by any event and the same program can be used by another process without loading such a re-entrant program a second time. System programs of operating systems are generally developed as re-entrant programs (like device drivers).

Evaluating the data structures of the processes and the OS needed till now, simple structures solve a lot of problems. Linked lists that connect tables containing pointers to other linked lists are all what is needed. This structure is not complicated as a structure but only as an object of dynamic change. The structure created in this way is a tree structure. As operating systems did not operate on trees (like e.g. object oriented programming or semantic language recognition) in the 1980s, the data structure 'tree' is not of interest in this context. 'Trees' are objects of mathematical graph theory. The origin of graph theory was laid by the British mathematician and lawyer Arthur Cayley (1857).

The performance of the system became an issue and so the job of the professional system performance analyst was born. One of the most crucial parameters of system performance was the structure and sizing of the paging file. We still understand bad disk management as a cause for poor performance. Defragmentation is one remedy for the home PC.

Hard Disk Drives

Paging was first used in research with magnetic drums. Those devices were slow, heavy and of limited storage capacity. The IBM 650, the first genuinely programmable computer presented in 1955, worked with magnetic drums. The IBM RAMAC 305 followed in 1958 with disks.[64] William Goddard and John Lynott invented the first hard disk drive at IBM in the 1950s at the Random Access Memory Accounting Machine (RAMAC) project for the IBM 305 RAMAC computer.

Fig. 11: Read / Write Heads of an Early Hard Disk Device

A hard disk consists of different layers of magnetic plates. Read/write heads write data in segments of fixed size on circles. So the heads move from one circle (cylinder) to the next if a head performs a read or write operation on any one of the layers. The information is recorded within the segments in bytes of fixed length. The first byte size used was 6 bit, we now use 8 bit with error detection and correction for one bit errors. The segments on hard disk media are addressed by the number of the layer, the number of the circle and the segment number within the circle. All circles store the same number of segments regardless of the size of their circumference.

From the beginning, paging files were laid in the middle of the disk in order to minimize the mean access time of the read/write heads to the circles and segments of the page files. If possible, specific page disks were used so the read/write head was generally in the disk area of the paging file.

If the segment size of the paging file is as long as the segment size of memory, the address of the page searched in the paging file can easily be calculated. An easy numbering scheme can be applied.

Job Management and Program Structure

Application programs are handled as a task, a process or a group of coordinated and interconnecting processes by the task scheduler. This was simplified with the job management. A job was introduced in operating systems as a technique to plan the sequence of application programs and their use of resources. If e.g. a batch job should be planned to print every night all invoices that 3 affiliates collected during the day on punched cards we could break that down to the following job steps:
(i) The objective is to sort the input cards with the amount to be billed in order to update the customer accounts. So the program could e.g. read each of the three stacks of punched cards in parallel from three different card readers and write the sorted data records on a tape. - Alternative strategies are possible if for example there is only one card reader.
(ii) The second step could be the comparison of the customer account records (may be also stored on tape) with the new invoices in order to update the customer accounts. The program writes the new customer account data on a tape.
As an example, if customers get monthly notes about their account and if these notes are distributed by the affiliates, the customers tape can be processed,

creating again three data tapes with one tape for each affiliate. Then each of the tapes is printed. The printed paper is also on a stack and sent to the affiliate.

To do this, a job management language was defined. This language could serialize the processing of application programs and reserve devices (like a card reader, a tape or a printer) for the job. Later, OS allowed that the job was broken down in job steps. The reservation of devices was valid only for one step. And later it was found necessary to synchronize different jobs, so one job could wait for the results of one or more other jobs. The job management language defined input /output channel numbers. The application programs could reference these input/output streams, e.g. through the channel numbers. This allowed dynamic allocation of devices to jobs.

The program that merges the above mentioned three input data streams from the affiliates might use Job Management Language and look like that:

```
// START JOB, JOB-OWNER=ACCOUNTANT
// FILE AFFILIATE1 CHANNEL=0 READER=R1
// FILE AFFILIATE2 CHANNEL=1 READER=R2
// FILE AFFILIATE3 CHANNEL=3 READER=R3
// PRGM MERGE_INVOICES
// END
```

The program MERGE_INVOICES merges the data (that are assumed to be in ascending order) and could work like that:

(1) The data structure of the input data record is defined in three buffer areas, each for one card reader, like: RECORD INVOICE NAME (CHAR30), AMOUNT (CURRENCY), DATE (DD, MM, YYYY)
(2) DO WHILE there is at least one card left in one of the card readers:
    a. The program reads from each of the card readers channel numbers 1-3.
    b. The program compares content of the three NAME fields.
    c. The record with the smallest NAME data field is put in the output file.
    d. The program re-reads from the reader that provided the smallest NAME data filed.
(3) The program closes all files and thus secures the data records.

(4) The program gives a statistical report for the job manager to check if all records were processed properly and none was forgotten or manipulated.

Soon, smart system analysts found out that SORT & MERGE are daily needed operations in data centres. To find and apply efficient sort and search strategies were topics on many seminars for application programmers. So the provision of one efficient program was worth a lot of money. A new market was borne, although it did not last for much longer than a decade.

Later, the channel reference number could be used to hide technical details of the devices from the application program. Only the channel number and the data structure had to be defined. Job managers and data centre operators found this a helpful abstraction.

In the first days, the programmer wrote his entire program in machine code or assembler language. The input and output instructions were part of the application program as well as all CPU management instructions. This was obviously no longer possible when the above mentioned computer model and application mix was used. Consequently, a supervisor program was developed that united all the management tasks and that created a virtual impression of a high level computer to the users. Task handling, peripheral device management, management of the batch jobs, and later of the time sharing users was the job of an OS. The reader perhaps already found that error management was a big issue till now. Now, the application programmers could concentrate on their application problems. This trend was enforced after 2000 with the programming environments. Examples are Visual Studios of Microsoft or the Java Development Kit (JDK).

Operating systems depended on the machine architecture, which – at first - was proprietary to every machine or family of machines. Later, standardized operating systems were developed that ran on the machines of different manufacturers and used different processor architectures. In the 1980s, the first machine-independent and portable programs could be coded.

## PROGRAM STRUCTURE

Application programs were structured according to the Turing and von Neumann concept as a sequence of procedural statements. One instruction (or programming statement) followed the other in the sequence defined by the programmer. Program loops, jumps to the end and conditional comparisons were permitted. To describe the logic, a kind of programming language will be used that is not fully formalized but offers the reader the necessary insight. This technique is called pseudo code.

The typical application program looked like that:

(1) START Main Program
(2) Definition of Data Variables and Constants
(3) SUBPROGRAM A (variables a1, …, an)
(4) SUBPROGRAM B (vaiables b1,…, bk)
(5) BEGIN
(6) Here comes the program body with all program statements, loops, conditional IF … THEN … statements, INPUT / OUTPUT statements etc
(7) END

These high-level programs were usually compiled and translated via object code in machine code. Some companies offered for special application areas (like controlling reactors) interpreters that interpreted the high level program and thus made sure that program errors could not stop the computer system on the condition of an error-free interpreter.

This structure was enhanced when the application programs dealt with asynchronous actions of the users of window oriented user interfaces, and object oriented data and program structures were introduced to increase programming productivity.

Since the beginning, operating systems needed a more complex program structure to handle a lot of asynchronous events in a real time environment. Events depend on the machine design so the main processor needs data structures that correspond to the operating systems data and program structure.

.

A simple example is provided by listing different events and the address pointer to the program module that should run, as soon as the event is signalled to the CPU.

| | | |
|---|---|---|
| Input/output event disk | → | Jump to address A1 |
| Input/output event tape | → | Jump to address A2 |
| Input/output event printer | → | Jump to address A3 |
| …. Further input/output events | | |
| Division by Zero | → | Jump to address B1 |
| Floating point overflow | → | Jump to address B2 … |

## 5. CUSTOMERS NEED PROGRESS

The Operating Systems (OS) of the general purpose computers became sophisticated masterpieces. Marketing's idea to bring a high level computing machine to the customer was commercially successful. The OS offered the customers commands like 'create file', 'stop job', 'reprint batch output'. These commands were translated in a series of machine codes. Millions of lines of code supported the customers in this way.

Customers needed more. The historic development is best documented in areas of academic interest and research. The health sector is a good example. A Patient Data Management System (PDMS) as outlined in the box of chapter 3 consists of modules for real time data processing and others for administration. When the integration of the earlier isolated system began, designers stated: "The numerous functionalities and management of different data types present the following design requirements: a) abstract definition of data types, b) unified view of data during the development phase, c) distinct levels of data management , d) minimum coupling of the system's modules and higher degree of system flexibility."[65]

The architects of this PDMS planned distributed systems that hold a large amount of data in memory and in different locations like laboratory data, bedside data in the Intensive Care Unit and administrative data in the hospital office. Of course, systems can fail. That was why they planned commercial transaction oriented systems (OS/2 Extended Edition, OS/2 Data Base Management) and could size the system to high performance. The data model is mostly relational although the authors talk about objects. A complete object-oriented approach was presented at the same symposium from Los Angeles. "Objects in this model are defined based upon more primitive objects. They are either versionable or non-versionable. Versionable objects are the collections of their non-versionable counterparts tagged by a time sequence …".[66] The importance of time and the processing of an enormous amount of big data were discussed from the beginning of integrated hospital management systems.

To meet today's goals, the HANA Technology is needed. In 2012, SAP and the Charité Hospital of Berlin presented at the CeBIT fair in Hannover support systems for medical diagnosis that analyse within milliseconds all medical records of the hospital and inform the physician real-time about cures used and effects produced in the past. The product is released for general application.[67]

Customers needed additional functionality like systems with zero downtime, fast data base and knowledge base systems with structured and unstructured data and Giga-Bits of data in milliseconds at their fingertips in any place and at any time.

## NETWORKS FROM THE 1970S TILL 2013

Readers interested in ancient communication techniques (with flags, mirrors and fire) may like the introductory remarks in "The Early History of Data Networks".[68] This book focuses on wired networks transmitting electric signals and especially networks with active elements like switches, routers and servers.

Conventional telephone networks differ substantially from computer networks in their behaviour of overloading. Classical telephone networks switch and hold lines between two or more people or a computer during their communication. These lines were initially physical lines but later consisted of a frequency or a data slot on a multi-user line. In any case, whenever the system was at the end of its capacity a further attempt to communicate resulted in the "busy" signal. Computers act differently as

Fig. 12: Accoustic Coupler

**Computer Networks** began in the 1970s with low speed point to point connections via telephone lines. Data was transmitted as 'speech" via acoustic couplers. This photo of an acoustic coupler is a courtesy of Don Anderson.

For multipoint connections the computer manufacturers invented proprietary protocols. Although governments tried to defined 'open' protocol suites like DECnet, Systems Network Architecture (SNA, IBM) the Internet protocols became today's

computers accept even in overload situation further requests for computing or communication. Computers then share the resources according to the priority and the service classes.

Today's networks are organized with computers so the overload logic applied is the way computers act. As an example, if on a modern cell phone network the system is overloaded, the SMS service will be given less priority by the network computers. The telephone communication gets more bandwidth and priority over SMS service.

Networks started in the 1970s as point to point connections from a general purpose computer to a place connected via a communication line. Early communication lines were telephone lines. Data were transformed into speech with an acoustic coupler for transmission. You could hear your data.

In the 1980s, mainframes controlled greater networks with data lines. Usual speed for applications was around 19.600 bit/second (baud). IBM mainframes with Multi Virtual Systems (MVS) used the exit programming in the Resource Access Control Facility (RACF) of MVS in order to program authentication and access restrictions. SNA was initially built as a collection of n star shaped networks with mainframes as important network nodes. This allowed the implementation of tough and client specific security measures.

Other computer manufacturers had comparable concepts as their mainframe was also in the centre of the network. Alternate routing in case of network failures or overload on specific transmission lines or servers was at first difficult and scarcely used. In the 1990s, routing in mainframe centred networks became more versatile as network computers became more sophisticated.

In the 1990s, "any-to-any" communication was the buzz word and the challenge of the networking experts. Soon the message was that "the network is the system". Empowerment of networking was possible with more powerful networking equipment (e.g. from CISCO). Access was controlled either by virtual private networks, by access control blocks in the headers of the data packages or by cryptography. Local Area Networks (LAN) worked like radio stations on wires: Data was sent to all devices connected to the LAN. These devices had (and still have) separate LAN addresses. Only the device accepted the data that had the media access address which was stored in the data package. Later LANs were broken into segments e.g. by routing firewalls.

.

The Internet technology allowed networks to search automatically via alternate paths for new connections.

Wide Area Networks in Europe started with 1 GBit/sec around 2000, 10 GBi/sec around 2010.

Characteristics of modern networking technology are routing with optical modules with speeds above 10 GBit/sec. The higher layers of the Internet are Web 2.0 or above. This provides interactive use of Internet sites, availability of social networks, Voice over Internet Protocol (VoIP) telephony and the use of various servers like Youtube and Dropbox.

SkyDrive[69] and other Cloud Services were already described. See figure 13 for a practical example.

Internet services of several companies became so comfortable that some people use only the service of one or a few companies. Especially some users of Facebook inform their friends that the times of e-mails have passed. Facebook and maybe Twitter is all a person seems to need, is the opinion of some users.

In January 2013, the French telephone Company Orange informed that Google pays for their service.[70] The political discussion about the Internet business, net neutrality and freedom of the Internet may become a further push from such news.

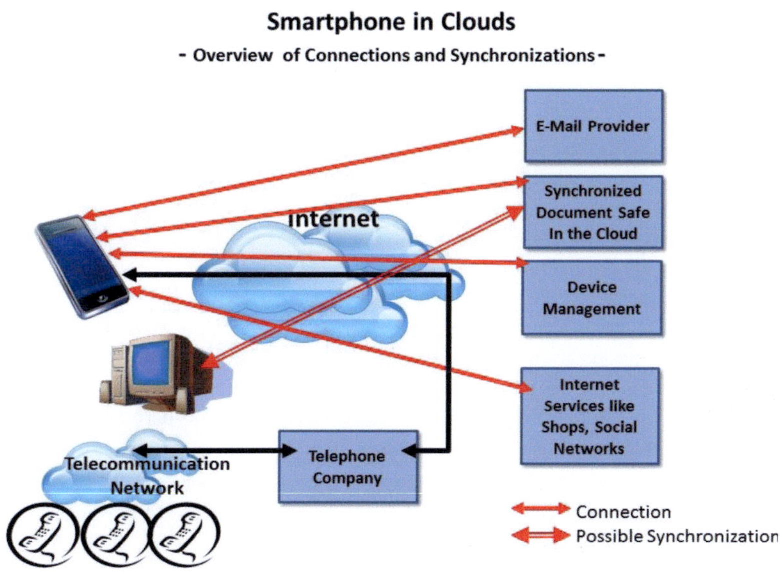


Fig. 13: Smartphone Functionality

In 2013, users take highly integrated infrastructures for granted. Fig. 13 explains the different networking connections that smartphones hold regularly. Special communication channels are possibly used (i) for e-mail, (ii) the management of the mobile device (like search function, update of the system software of the smartphone, automatic deletion of the private content of lost smartphones), (iii) access to all functions of the Internet, (iv) access to a document safe in the cloud that provides optional synchronization depending on the settings given by the user to the documents, and (v) access to the telecommunication networks. The document safe may contain documents, e-books, digital pictures or any other digital content. Other possible functions like App stores of smartphone vendors and program-to-program communication functionality are not described in figure 13.

In 2013, the slogan "The Network is the System" became reality by using cloud computing technology. Providers of integrated solutions, as described in figure

13, are e.g. Microsoft, Google and Apple. The different companies used individual App release and billing concepts in 2013.

## PROGRAMMING LANGUAGES AND TOOLS

Assembler programmers are in the programmers' community the experts with most freedom. They can define any place in memory as an instruction, or a data field, or both at different times and – in errors – at the same time. Careless use of this freedom could become a problem. The computers used to run wild. The CPU stopped at undefined memory locations. As memory dumps were not always easy to print or to understand with a CPU that ran wild about 20.000 instructions ago, an instruction by instruction check of the assembler program with hands on the machine were carried out (almost daily). Productivity was awfully low. Programming was considered good work when the programmer managed to code three to five instructions per day (measured from start to end of a programming task) as an average.

High level application programming languages were needed and new ideas were borne by researchers. Pioneers of programming were among others Alan Turing, and in the end of the 1950s John Backus and Peter Naur for the Algorithmic Language ALGOL 60. The idea was to translate a high level programming language with a program called "compiler". The "high level" should ensure higher productivity. The result of the translation was object code that could easily be loaded into the computer by the "loader" of the operating systems and processed by the CPU. Object code was basically machine code (but with relative address notation) that could be loaded in almost any place of the main memory.

Hoare had this background when he started to define further objectives for programming languages. His first objective was 'security'. Hoare reports:

> *The principle that every incorrect program should be rejected by the compiler and that every syntactically correct program should give a result or an error message that was predictable and comprehensible in terms of the source-language program itself. Thus no core dumps should ever be necessary. It was logically impossible for any source-language program to cause the computer to run wild, either at compile time or at run time.*"[71]

Hoare's second principle was 'brevity of the object code produced by the computer'. The intention was to reduce memory space. This and the third of Hoare's principles state that all has to be done in the most efficient way. Although memory space is cheaper now, Hoare's principles are still valid from this general point of view. The fourth principle was that the compiler should do a compilation in one pass. This was important as it motivated to study languages with their formal and semantic aspects. So ideas of computer science and linguistic research[72] met. Some applications (like the control of nuclear power plants) needed application programs with zero downtime. One possibility was to use interpreters instead of compilers. Interpreters do not translate a high level programming in object code. Interpreters never give the control of the computer away to application programmers and interpret all programming statements. In case of errors (like a division with "zero"), the interpreter still controls the computer system.

To construct a compiler or interpreter, a specification was needed. To simplify the task, the grammar of the programming language was a context free grammar. It took several years till a formal specification language for programming languages was invented. Later it was called Backus – Naur - Form (BNF).

BNF notation is

<symbol> ::= <expression>

The symbol " | " is used to denote an "or" in the logical meaning of "vel", not "aut".

To give an example one valid statement is

<integer> ::= 1 | 2 | 3 | 4 | 5 | 6 | 7 | 8 | 9 | 0 | <integer>

The definition can be applied several times (i.e. recursively). So we get the integers "1", "01" and all others.

BNF may be used to define a train by

<train> ::= <engine> | <train> <waggon>

If we give a specific symbol for <engine> and <waggon> we can derive all finite series of this engine with n waggons. Trains with multiple engines are not trains according to this definition.

This formal definition was a breakthrough for the specification of context-free programming languages. BNF was easy to use and to understand which was not true for other methods. BNF is no escape from Gödels proof in mathematical logic saying that these subjects cannot be solved in a non-contradictory way. Hoarse first objective can only be met in the sense of a better approximation, never for 100%.[73]

Backus worked for some time as an employee of IBM. The value of his work was not directly applicable in the area of commercial programming. Some researchers of IBM worked on programming languages that came near to the mentality and computing problems of accountants. Documented in its first version in 1959, the Report Program Generator (RPG) allows the easy construction of programs for commercial bookkeeping and billing. These applications require mostly lists with footers or headers containing sums, printouts like bills, salary notes etc. So the inherent programming logic of RPG is the processing cycle.

As a general purpose language for business needs, the Common Business-Oriented Language (COBOL) became the language of choice for most commercial programmers in the 1970s till 2000. COBOL was discussed in the CODASYL forum and first versions come from the late 1950s. The American National Standards Institute (ANSI) defined COBOL in 1968 as ANS COBOL. This version was used in the industry for many years as an accepted way of documentation and a portable language.

Programmers used to have a personal collection of sub-programs and of data definitions (macros). Collections of 'sub-routines' (like the sorting of data arrays or the definitions of an address, bill or an employee file) were often helpful for the employee and the employer. Such collections improved productivity. Recall that programmers' productivity was about 3 to 5, maximum 10, lines of code per work day and per programmer.

Here, an example is provided to demonstrate the usability of this way to work.

```
DATA DIVISION.
FILE SECTION.
...
FD  Customers.
01 Customer
   02  Customer-Number         PIC 9(7).
   02  Customer-Name           PIC X(30).
   02  C-Fee-Owed              PIC 9(4).
   02  C-Amount-Paid           PIC 9(4).
```

That sounds hopeful as practical experience was compliant with Gödel's prediction. Dijkstra notes:

> *"Recalling that our true concern is with really large programs, we observe as an aside that the size itself requires a high confidence level for the individual program components. If the chance of correctness of an individual component equals p, the chance of correctness of a whole program, composed of N such components, is something like*
> $p = p N.$
> *As N will be very large, p should be very, very close to 1 if we desire P to differ significantly from zero!"*[74]

A remedy by mere formal definition of application programming languages or by other functions of an operating system was not in sight. So companies produced test tools that printed helpful overviews like e.g. a statistic of program tests done with test files. The program printout included not only the programming instructions. How often DO WHILEs, IFs, GOTOs etc were processed was also listed.

The author remembers test situations where parts of the application programs were never reached by the test data. After some research, the programmers found the solution. The clauses of the legal base, that made programmers code these modules of the program, were logically impossible. So politics had – with much dispute - passed a law, and it was contradictory. Gödel was right, again. What a nice way to improve political decision making with ICT.

Dijkstras problem with big programs was a substantial burden for database companies like Oracle, Software AG, IBM, HP. The customers needed more and more integration. The complexity of the programs rose. Why? People in the information society became used to computers. They did no longer accept simple solutions. People now wanted 'more justice'. More justice made contracts and laws more complex. More detailed cases had to be considered. People also argued that computers should be able to find the data automatically and provide more customer friendly solutions. Previously, agencies and companies had asked their customers to provide these data themselves.

At the same time, database systems were rolled out to companies and agencies. One solution, that was provided in the early 1980s, was Computer Aided

Software Engineering (CASE). Software AG, the owner of PREDICT, informs in its 2010 version what these systems do:

*"The following types of information can be stored and maintained:*

- *What data is stored and where. This data reflects the information structure of an organization.*
- *Which processing objects work with the data. This data reflects the functional structure of an*
- *organization.*
- *Who creates data and/or objects that process the data. This data reflects the personnel working*
- *with the data processing system."*[75]

In the 1970s, Jean-Dominique Warnier, a french computer scientist working for 'Compagnie des Machines Bull', invented a brilliant method to structure programs. From an empirical point of view, a program does nothing but changing input data to output data. The simplest way to do this transformation should be the application program.[76] Warnier described easy ways to get a program structure by evaluating the structure of input and output data. A comparable method was presented by Michael Jackson in the U.S. He supported his method as a product in the early 1970s.[77] Both methods became popular as the basic elements were graphics that programmers knew from flowcharts. Both methods had the advantage that the application programmers hardly had a decision to make. The structure of the program could be derived semi-automatically through recipes.

These methods were predecessors of Computer Aided Software Engineering (CASE) for relational data base management systems and of object oriented programming. The methods of Warnier and Jackson were of great help for many application programming teams. The weaknesses were insufficient theoretic foundations. For bigger application systems, no acceptable and forward-looking automatic graph builders were supplied for the Jackson Method and for Warnier's Method. - Relational database management systems were better suited for automatic application generation. Oracle delivered with Oracle Forms a product that allowed the automatic construction of dialogs for online-update applications based on the table structure of the database. This was later enhanced to a CASE environment (see Fig. 17).

.

Commercial computing is the processing of large amounts of structured data. So programming languages for commercial data processing needed easy to use data definition and file management abilities. The next table tries to describe the development, although there are different ways to describe progress made in data and file management.

| Year (estimated by the author) | File Management (according to Kevin M. Whitney)[78] | Database and Transaction Management | Remarks |
|---|---|---|---|
| 1940 - 1970 | Each programmer wrote his own input/output routines | Input/output on Teletypewriter (TTY) or other displays was not secured. | For cell phone, μP programing and other appliances this technique is still used in 2013. |
| 1950 - 1975 | Access methods were<br>• file catalogue management<br>• sequential files fixed and variable record size<br>• direct access files with fixed record size<br>• index sequential file management (ISAM). | Simple transactions were programmed with individual techniques.<br><br>Some computer manufacturers provided one step transactions with ISAM. | ISAM files were used with primary and secondary keys. To provide acceptable performance and response times system administration had to understand the use and distribution of the keys in te records. |
| 1960-1980 | Several files were interconnected with identification systems invented by the application programmers. | These semantic entities were not secured. | The emphasis was on report generators and retrieval systems like e.g. EASYTRIEVE. |
| 1980-2000 | IBM began delivering systems that integrated a) transaction processing and b) data management in transactions with e.g. IMS/2. | The application systems began to allow online read from databases and data entry through special data entry systems. Soon online update with secured transactions followed. | The report generation was still important. Data warehouse technology was invented to reduce the performance burden from the applications with online update functions. |

| | | | |
|---|---|---|---|
| 1978-2000 | CODASYL Database concept was used. | The concept of transactions was understood and stepwise applied. | |
| 1980 – today<br><br>2000 - today | RDBMS were used<br><br>Object Oriented RDBMS are used | Customers found it self-evident that RDBMS are integrated systems. Especially Oracle soon provided its database system with integrated application development tools. | Some manufacturers created machines that operated on relational databases only. One was Intelligent Database Machine (IDM). |
| 2000 – today<br><br>2010 - today | Graphical User Programming was used.<br><br>Apps are used in Smartphones. Touch screens are normally used. | User programming started as customizing by users. When it became a standard and secure development technique it laid the basis for Cloud Computing. | Navision was one of the first companies that provided their individual users of accounting software enormous online flexibility to change data records in a running system. Internet providers soon allowed their customers to create their own internet site with graphical programming. |

A transaction is the action of an online user between the input of data and the response from the machine (one step transaction). Soon applications needed multistep transactions. When we buy a product in the internet, a multistep transaction is performed: (i) The credit card is checked. (ii) The stock of the product we want is checked. (iii) The product is reserved. (iv) Delivery is planned. (v) The costs are documented and registered by the credit card company. If one of the systems used fails, a restart of the transaction is necessary. Nevertheless, the internet customer should see a smoothly operating integrated system only.

What COBOL programmers were looking for, when they collected their little examples of program code and data definitions, became possible with object

.

oriented programming. This programming technique provided a systematic approach for productivity increases:

(i) Objects were defined that had an internal and external data structure. The internal data structure and the internal code are hidden from external access. Objects communicate via well-defined methods with other objects.

(ii) Classes are defined, and sub-classes are possible. The programming environment provides an enormous set of classes to the programmers. These classes define among other things the output possibilities to windows oriented surfaces on displays of any kind. The productivity bottleneck of conventional programming languages could be eased.

(iii) An inheritance technique allows defining processing modules that are valid for an object and all defined sub-objects.

Object oriented languages were a significant improvement. The transactions on the databases are not easy to define in an object oriented way. Many problems translate easy into RDBMS structures but not in object oriented database systems. Some software providers delivered XML database systems. Today, Java (in 2013 a product of Oracle Corp.) is a programming standard in the open source programmers' community.

When the widespread use of mobile devices asked for applications, the software development was challenged. Icons on the surface of the mobile devices allowed starting an application with a tip of the finger. That made computing at the finger tips, an advertisement slogan of the late 1990s, become true. These application were called Apps.

The providers of operating systems for the mobile devices provided software development tools working on an object oriented model and based on enormous module libraries to speed up programming. The chapter on highly integrated software describes the challenges and the need for better software quality.

The next page shows the desk of software developers working with Eclipse.

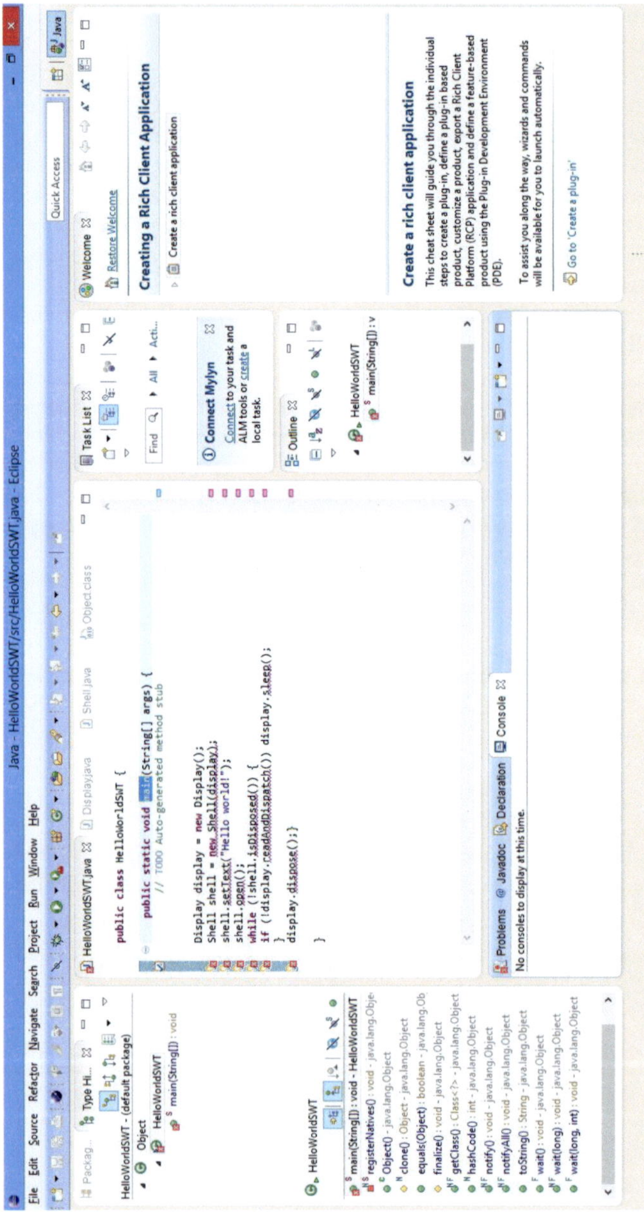
Java - HelloWorldSWT/src/HelloWorldSWT.java - Eclipse
File Edit Source Refactor Navigate Search Project Run Window Help
HelloWorldSWT.java
Display.java
Shell.java
Object.class
public class HelloWorldSWT {
public static void main(String[] args) {
// TODO Auto-generated method stub
Display display = new Display();
Shell shell = new Shell(display);
shell.setText("Hello world!");
shell.open();
while (!shell.isDisposed()) {
if (!display.readAndDispatch()) display.sleep();
}
display.dispose();}
}
Task List
Connect Mylyn
Connect to your task and ALM tools or create a local task.
Outline
Welcome
Restore Welcome
Creating a Rich Client Application
Create a rich client application
This cheat sheet will guide you through the individual steps to create a plug-in, define a plug-in based product, customize a product, export a Rich Client Platform (RCP) application and define a feature-based product using the Plug-in Development Environment (PDE).
To assist you along the way, wizards and commands will be available for you to launch automatically.
Go to 'Create a plug-in'
Problems
Javadoc
Declaration
Console
No consoles to display at this time.
Quick Access

## DATABASE AND TRANSACTION SYSTEMS FROM THE 1970S TILL 2000

The processing of a sequence of jobs or of a high number of parallel and synchronised jobs was clearly insufficient to handle hospital management systems. The problems were that data were scattered throughout the various systems, often denoted with the same identifier but with different meaning. NAME could mean for example the patient, the relative of the patient to be contacted in case of an emergency, the person to be addressed in the insurance, the physician who documented a diagnosis. This may astonish but these were daily problems of coordination of programming teams of individual behaviour and intentions.

In the 1970s, smaller organizations (less than about 2.000 people employed) documented their data and program tapes often in an individual way. Larger companies used the first automated – often self-programmed – file, tape and media management systems. Data centres could not even hope for robot systems like we use them today.

Another problem of the time was the correct synchronization of data of the same processing cycle at a specific date. Batch files were numerous in those systems, and documentation was sometimes incomplete. So the operators of data centres might get confused, and sometimes they used files that did not belong to the same time cycle.

Such problems were anticipated early. The Conference on Data System Languages (CODASYL) was formed in 1958 to attempt to standardize computing languages. It found it impossible to get acceptance of any language. In the late 1970s, CODASYL proposed a concept for data base management and its use with programming languages.[79]

The principles of data base management systems are:
a) keep data persistent, which means defined in syntax, semantic and time,
b) avoid redundancy, which means that every single data is stored on one place and only on this place so that all data will be shared by all application programs,
c) maintain data integrity, which means that all data in the data base are complete and correct and link to correct data in other areas of the data base,

.

d) control access properly,

e) handle processing conflicts at one place.[80]

Consistency of a data base is violated if one transaction transforms a data base with full integrity to a data base without full integrity. It could be expensive and difficult to repair an inconsistent data base.

When those responsible for the organization of an enterprise understood the costs and potentials of computing, they requested to get the same data model throughout the enterprise. One major objective of data base systems became the portable data definition. It should work for the database used for the administration and in the same way for the database software used in research and development.

A systematic approach used three languages with data base systems: The Data Definition Language (DDL), the Data Manipulation Language (DML) to define the operations and instructions given to the application programmers in the programming languages and the Storage Structure Language (SSL). SSL is for the database administrators to run a database with high security, tolerable performance and high flexibility on all reasonable disk, CPU and networking configurations.

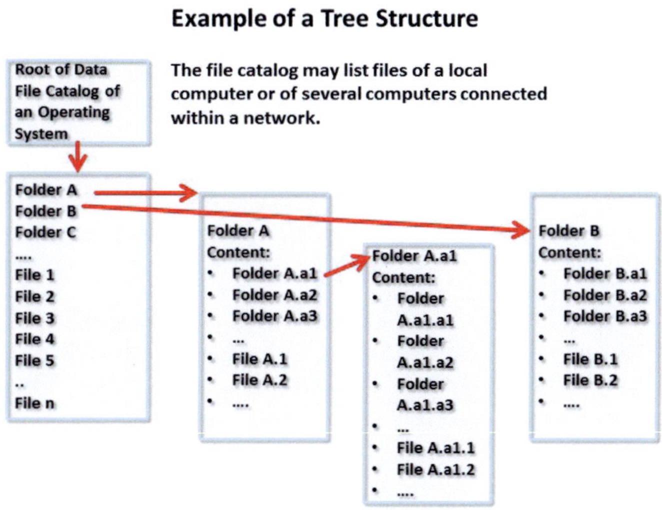


Fig.14: Example of a tree structure

We know from physics and mathematics that notations are a powerful way to simplify understanding of the world. This is also true for databases and the notation to describe data dependencies. In the 1960s, two aspects were understood. First, Charles Sanders Peirce (1839-1914), one of the founders of the philosophy of pragmatism, saw a great difference between a conception and the pragmatic effects. This maxim of pragmatism translates in database theory to the definition of metadata (as data about data) and the instances of data defined in metadata. As an example, the macro definitions discussed above describe the structure of data in the sense of metadata. Actual data ("Charles Sanders Peirce" in the "NAME" field) are instances (or: instantiations). That data description concept of data bases allows helpful classifications in other areas like knowledge management, open data portals or myth theory[81].

Second, the basis of early database design is graph theory. The data structures are not so numerous:

- tables, which are sequential lists with possibly sub-lists
- linked lists, doubly linked lists
- indexed tables
- trees
- networks, which we usually break down in simpler data structures.

If this notation is applied to a set of data records for “car” and “tire”, the data records hold the following relation:

- Between the metadata “car” and “tires”, there is a 1:n relation as a car is equipped with 3, 4 or – with the spare tire – 5 tires. An instance of a car is a “Ford Mondeo”. Depending on what processing is wanted, “Ford Mondeo” is either the name of a definition of metadata (to be used within the Ford Motor Company to produce the Mondeos) or an instance of the metadata “car”. A garage management might use the second choice and look at cars rather superficially. They might use “Ford Mondeo” as a simple label different from “Volkswagen Golf” and others. They don’t bother about more. Ford Motor Company produces many different “Ford Mondeo” so the Ford Motor Company may take “Ford Mondeo” as a name for metadata with thousands of different instances.
  The database task group of CODASYL and some database definition languages use the term “occurrence” instead of “instance”. Of course, every software manufacturer creates a special language for his product. The 1:n - relation is also called an owner-member relation.
- Another kind of relation between metadata exists between the metadata “teacher” and “pupil”. As each teacher is connected to n pupils and each pupil to m teachers, this relation is called n:m - relation. “Joe”, “Claire” are instances of the metadata “pupils”.

To make the database design complete, the data base design needs a notation for the internal structure of the metadata. Metadata are a data description method for describing data base records and data fields.

To describe one example, pseudo data definition language is used. Thus, the necessary insight is given without worrying about the formalities of programming languages. The example uses for “teachers” and “pupils” the same data record structure “person”.
“/*” indicates the rest of the line as programmers’ comment.

```
SET IS TEACHER
    MEMBER IS PERSON, DUPLICATES NOT ALLOWED FOR
    PERSON-ID.

SET IS PUPIL
    MEMBER IS PERSON, DUPLICATES NOT ALLOWED FOR
    PERSON-ID

RECORD IS PERSON.
    PERSON-ID INTEGER
                /* One might use "CHARACTER" and store only digits.
                /* Using automatic data field checks is better.
    LAST-NAME CHARACTER (30)
    CHRISTIAN-NAME-1 CHARACTER (20)
    CHRISTIAN-NAME-2 CHARACTER (1)
    STREET CHARACTER (35)
    ....
```

To tell the database system that there is no metadata with a higher hierarchy than "TEACHER" and "PUPIL" the notation of CODASYL requires and additional remark "OWNER IS SYSTEM".

The application programs use these definitions within the data definition section. If the database functionality is not sufficient, links may be added and many more ideas can be programmed. Database definitions and the definition segments of programming languages fit together automatically.

Software teams and companies use thousands of such definitions. They prefer professional management of the metadata structures. Products like PREDICT© help in the form of an active data dictionary that is used by database machines and by programming languages. An integrated view and handling of data is possible in a highly productive way.

The disadvantage of the early CODASYL notation was "that external and internal model design parameters are intermingled in the DDL and are interdependent".[82] The CODASYL 1978 DDL provided the systematic and reliable definition of the CODASYL Data Base Standard.

At the same time, IBM tried in San Jose an alternative with the Project R. The theoretical basis was the work of Codd on relational database systems which are called SQL databases. When in 1977 the results of Project R were available, a

.

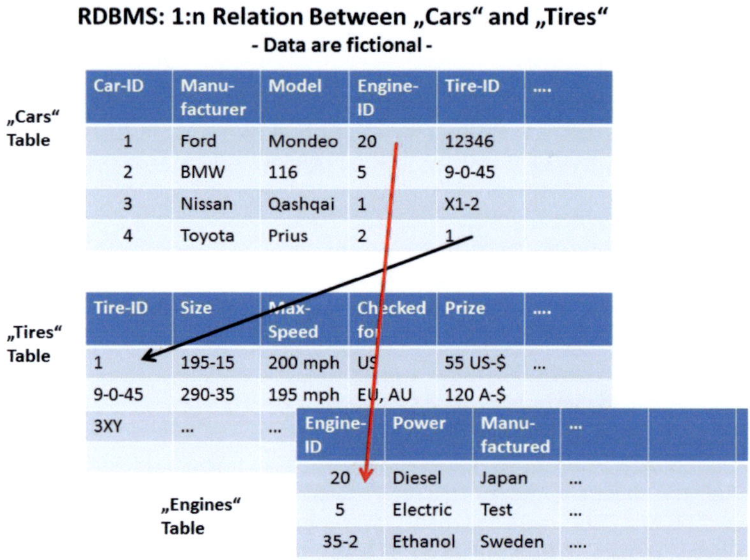


Fig. 15: Example of a Database Design with RDBMS

number of companies were founded. Oracle was one of these (see more in the table of the pioneers: Larry Ellison). At the end of the 1970s, Oracle and others offered the first version of relational database systems. At about the same time, IBM announced SQL databases SQL/DS for its DOS/VSE and VM operating systems and DB/2 for MVS operating systems.[83]

Relational database management systems (RDBMS) map more of the semantic aspects of the data structures in the formal database design. The example of the 1:n-relation between "car" and "tire" will show that RDBMS increase the flexibility of changing or enhancing database definitions as well.

RDBMS use tables to store data records and to define the dependencies between data records. The columns of the tables are called "attributes". Each table needs a unique identifier. Integers are often used as identifiers. This might produce the false impression of simple design construction. In practical problems, the search for useful identifiers might be difficult.

Tables with 1:n - relations are easy to understand in an RDBMS design. Academic training is not required. This is not quite true for n:m-relations. N:m-relations are mapped by using a table for the first data record, another table for the second data record and a third table for the n:m-mapping. Figure 16 with the "teachers" – "pupils" relations demonstrate this.

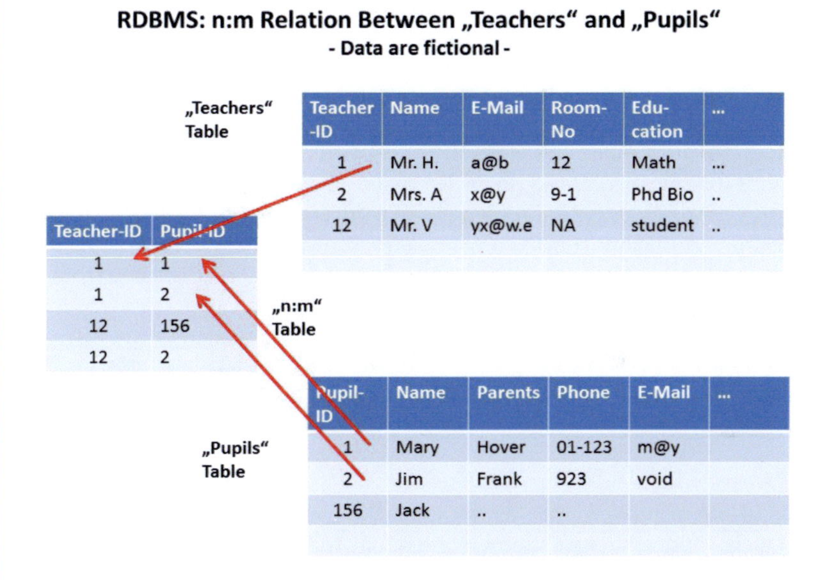


Fig. 16: Example of n:m - Relation "Teachers" : "Pupil" in RDBMS

Some peculiarities must be observed. The data base design with all tables must satisfy the normalization rules of RDBMS. Usually, inexperienced programmers build RDBMS applications with tables where identical content (instances seen as a line) occurs. The programmer had the idea that e.g. for every tire "size 195-50" in the warehouse there should be one line in the "tires" table. This idea does not work. Data manipulation rules like DELETE could only insufficiently be defined.

The normalization rules for RDBMS are the following[84]:

.

(i) General requirement: Ensure that all entities are uniquely identifiable by a combination of attributes and/or relationships.
(ii) First Normal Form: Remove repeated attributes or groups of Attributes.
Second Normal Form: Remove attributes dependent on only part of the unique identifier.
(iii) Third Normal Form: Remove attributes dependent on attributes that are not part of the unique identifier.

Using RDBMS in the first decade, data centres had a lot of problems with this technology. These databases are extremely flexible, and the application programmers used this flexibility. As a consequence, the table space (amount of disk space used for a table) on the disk was growing in areas not foreseen by the data base and system administrators. Suddenly, disks were full. Table space needed re-organization. The tables were re-distributed on the disks of the computer system. Often, new disks were needed. That required new strategies for nightly backup processing in order to provide a secured system the next morning. Often, the night hours were not sufficient for the batch job running for backup and update. If system administration was not able to address these performance problems in time, users complained about slow response from the applications.

Relational data bases allow integrating programs into the data base structure and logic. They ensure that the data are conforming to formal checks like "There is no car without tires."

Coming back to the subject of productivity, high level programming languages and RDBMS improved the quality and the productivity of programming enormously. Application programmers without formal academic training found it difficult to position the tables on the user screens in a way that the normalization rules and the database integrated checks worked well without testing. Oracle and other data base companies addressed this problem and offered rapid prototyping tools to the programming teams.

The following picture shows the first ideas of rapid prototyping used with Oracle Forms. The example uses the 1:n - relation showed above. The programmer refined the automatically generated screen and added data from other tables. The automatically generated application is consistent with the data base structure. If further data fields are added it is the programmer's responsibility to ensure consistency with the data base structure and with the routines programmed and integrated into the data base.

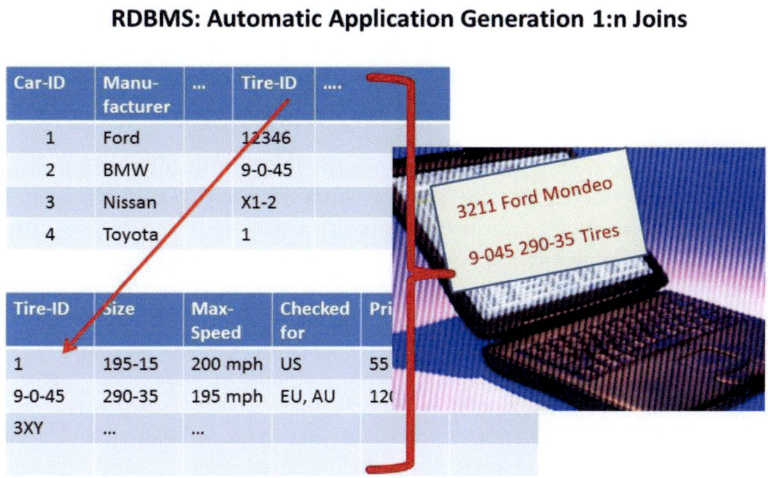


Fig. 17: Automatic Mapping of 1:n - Relations to the Application's Unser Interface on a Screen

These were first indicators that quality had to be further improved. In the 1990s, the installation of a new version of application or system programs was no longer possible whenever the specialists found a problem. Version planning was necessary. Weekends were used and back-up had to be prepared properly. System administration became an art. "Zero administration" and "zero failure" were anticipated by the customers. Companies like Microsoft and IBM designed their product portfolio according to such strategies. These companies were among the main drivers towards better system administration and reliable integrated systems.

## 3-TIER ARCHITECTURE STARTED 1980 - 1990

The high performance of modern Internet servers and Cloud Computing Data Centres is not only due to faster CPU's but also to a change of the computing architecture. The 3-tier architecture followed the general purpose computer architecture. Terminals became client computers with standardized dialog processing. The client was either a "fat client" (i.e. a fully functional PC with Windows) or a "thin client" with terminal server support from Microsoft or Citrix.

In the 1970s and 1980s, Teletype Writers (TTY) and then Cathode Ray Tubes (CRT) were connected to the general purpose computers with ever more difficult protocols and controls for optical presentation (e.g. highlighted data fields on the screen in various colours). This produced a heavy load to the CPU. An analysis of the code path of a CRT dialog was advisable and delivered the following results (example):

a. Processing of the input data of a dialog: 150.000 instructions
b. Productive processing of the application: 200.000 instruction, containing about 50.000 instructions for the file management, 50.000 instructions for the content of the dialog's data and 100.000 instructions for help messages and error handling
c. Processing of the output data for the dialog: 150.000 instructions.

Summing these findings up the result is that a standardization of CRT control and dialog management would reduce the code necessary to process interactive dialogs. When the PC was invented and became a cheap computer component in the mid 1990s, the computer architects naturally created a standardized dialog management (with several windows oriented techniques) that they processed in the PC. This enabled the computer industry to reduce the cost of computing. Looking back from today, historians are astonished to note that some computer companies did not see this development. The way to windows oriented user interfaces with a PC as a client was, from a technological and financial point of view, a logical way to go.

The 3-tier architecture became fashionable after 2000. When data base management systems were applied in more and more complex applications and when such application systems needed remote connections (like e.g. in e-commerce a connection to the bank to check the credit card), the separation of the sub-systems processing computer dialogs was an obvious idea. The separation was within the general purpose computer or – in the case of bigger systems – realized by several specific servers.

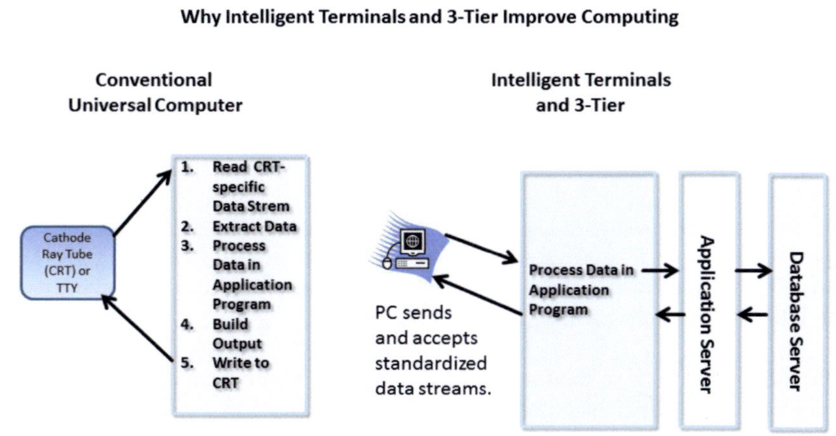


Fig.18: Computing Architecture for Commercial High Speed Computing

## WINDOW INTERFACE NEEDS OBJECT ORIENTATION

Object oriented languages were conceived in the 1950s and 1960s with the early versions of programming languages like SIMULA and LISP. There are several reasons why object orientation was needed:

(i) The note on 3-tier architecture explained the technical need for standardized user interfaces. When application programs became more complex, the users needed more flexibility for the use of these highly integrated applications. One example is the processing of a tax declaration. Tax officers need a flexible computer interface. They possibly start with entering the data of the bank account. Then they check taxpayer's information concerning the rent of a house. They may now go back to the bank account and check how much money the children received as monthly allowances. Such user interface flexibility needed windows. Each window stands for one aspect of a case. As the tax officer may get a phone call from another tax payer while working on one case, the user expects the windows of the different cases to be somehow connected and clustered to each case. Each cluster might have its own colour or other visual characteristic.

(ii) Programming such an application would be faster if the programmer could pass data attributes or settings of programming variables of the main window on to the sub-windows. The object oriented programming languages allow such an inheritance mechanism.

(iii) Software testing requires a lot of test cases and time. To increase programmer productivity, programmers of languages like COBOL kept their own sub-programs and data definitions ('macro'). Object oriented languages support such intentions by design. Sub-routines are encapsulated and called objects. Objects possess well-defined data interfaces and functionalities (called 'methods').

On the one hand, object oriented programming became more complex as programmers learned not only a language but a library of useful modules as well. Nevertheless, the productivity of well-trained programmers for commercial applications rose. Object orientation is usually not used for simple programming tasks or scientific calculations.

Object oriented databases are often used with object oriented programming languages.

An even more important consequence of object oriented programming was the possibility to program more complex application systems. Primarily, software companies intended to deliver application systems that the user can customize without professional programming skills. There was no place for computer science theory like normalization rules. The solution is graphical programming with limited freedom for the customers. Examples provide tool sets for Internet users allowing everyone to create one's own Internet site by just clicking on functionalities desired. Another example is the customization of an accounting system for smaller enterprises. Every accountant can develop the system needed for the specific purposes of the enterprise.

Of course, clients world-wide were happy with such systems that offered them for the first time the flexibility of instant software development. The application is ready for use without further thoughts about configuration and change management.

Early in 2000, data centres like those of IBM or Microsoft with several thousand of such clients offered step by step "computing on demand". The monthly billing is according to the computing and storage resources used.

To better promote this new functionality it was called "Cloud Computing". Three basic characteristics define a service as a Cloud Computing Service:

- The software offered allows the users an individual and easy customization.
- The computing resources needed are scaled real-time according to demand. As a security against denial of service attacks or erroneous program loops, the service providers usually request a special confirmation for higher levels of resource usage. Details are found in the draft contracts and in the offers of the cloud computing providers in the Internet.
- The billing is variable according to the resources used. So start-up companies can offer new services without risking a high amount of fixed costs.

Cloud computing does not necessarily mean that the services are used through the Internet. Private Clouds are possible for applications that need more security.

Cloud Computing is standardized by NIST[85]. Chapters 7 and 8 go into more details.

## SPECIAL STANDARDS FOR PUBLIC ADMINISTRATION

Public administrations in the U.S. and later in Europe defined their own standards. The main reasons are (i) to simplify the use of new technologies, (ii) to reduce costs by standardization, (iii) to ensure a secure service, and (iv) gain fair and non-discriminative call for tenders. Being a strong buyer in the ICT market, some European public agencies intend to push technological development of European industry by long lasting projects and contracts. There is a general belief that an example was set in the U.S. where government orders (e.g. from the Department of Defence) resulted in positive impulses to technologies. In the USA, Open System Interconnection (OSI) and other standardization activities were successful, at least at the beginning.

After about 2000, the work of the 'open' communities in the Internet, in Research and Development became more successful. One reason is that licences for open systems were published, practical and used on a large scale.

Standardization is helpful for internal planning of public administrations. Public procurement regulation accepts such standards only if no unfair restrictions of bidders occur as a result. But this is often the case.

Standardization of public administrations is usually slow. In the public area too many people and institutions take part in the coordination needed. This is one of the reasons why standardization in the public administration is often not successful.

Some standardization examples, taken from Germany and Europe, may help to see the difficulties. Better ways of coordination of ICT should be chosen in the future.

When in the late 1970s Virtual Private Networks were invented with packet switching capabilities, the public administration defined with the help of standardization organizations a suite of X-definitions. X.25 was the packet switching, X.400 was the mail system and X.500 was the directory service belonging to the mail system. This was not totally unreasonable at a time when IBM, Siemens, DEC, HP and others offered their proprietary networking and mailing solutions that were mostly incompatible. The intention to invite a large number of companies to deliver an offer in a call for tender was not successfully put into practice. First, the public administration had to pay for the development of solutions according to these standards. The ICT companies saw no reason to offer such services in a time when the Internet was clearly coming on the horizon. Second, the ICT service was extraordinarily expensive on the basis of these standards. Manufactures could hardly expect a return on invest through future call for tenders. At the time, manufacturers developed first Internet products for the ICT market. Third, the quality of this software for a small market was not inspiring.

When in the 1980s databases and transaction oriented systems were introduced into the market, the German public administration defined specific interfaces for transaction systems and databases. These interfaces were called Universelle Daten Communication Schnittstelle (UDCS) and Universelle Daten Bank Schnittstelle (UDBS). The standards were rarely used by the administration.

The most expensive adventure into standardization was undergone when the European Commission started in the 1980s the initiative to work out a "European Procurement Handbook on Open Systems (EPHOS)[86]". This work was accompanied by a European Directive obliging all member states to use only Open Systems in ICT applications of administrations. The success of the Internet soon accelerated technological progress in a way never seen and expected before. In 1995/1996 EPHOS was stopped. Complex systems were substituted by much simpler solutions. The public administration could cut ICT costs, and the ICT companies observed a new level of competition.

The German federal government still publishes ICT standards, called SAGA[87], upon request of industry and many agencies.

## PROJECT MANAGEMENT AND IMPORTANCE OF TRANSFORMATION

In the 1960s, computer projects drifted from research to engineering. People did not understand the rules that apply to project management and expected a project manager to be like a magician. Management set the project manager a triangle determined by budget limits, time limits and a Pandora of always growing functions and moving targets as a framework. Then he was expected to be successful. This was in most cases impossible. Early projects ended often "out of budget" and "out of time".

Many lessons were learned, and many remedies tried. Engineers learned to specify and plan a project completely before starting. Methodologies were found. Later in the 1990s, software development tools allowed rapid prototyping. After 2000, Electronic Data Systems (EDS) began with agile project management. By the way, the fact that EDS was later integrated into HP does not mean that agile project management is inefficient.

.

**1993: IBM Starts Transformation**
Success may become malediction. If the ability to adapt an organization constantly to new situations is reduced by past success, a critical situation comes up. IBM employees in Germany felt this in 1993. The company was in danger. A drastic transformation was necessary. The question was how and in what direction?

After the employees of IBM Germany had changed the company, they wrote a book about it with the title "Ändere das Spiel. Die Transformation der IBM in Deutschland und was wir daraus lernen können." (Murmann Verlag, 2010, ISBN 978-3-86774-109-5). Some of the lessons learned are reported in the main text.

The most important lesson to be learned in project management is that moving targets spoil every project. There is no alternative to knowing what is wanted. Moving targets are rarely successful. Computer Scientists had to learn those lessons often in a hard way.

Why do projects fail?

*Scenario 1:* The staff and the know-how needed for the project are not available. So the project manager plans without calculating loss of productive work time through sickness, vacation and job rotation. It is needless to say that the end is sad.

*Scenario 2:* The budget is not available. So the project manager starts with the assurance that he will be promoted. Another sad end is for sure.

*Scenario 3:* The project manager loses the support of the management as he works in silence, professionally and successfully. Suddenly, the project manager finds out that bad news is better for his career. All his colleagues with trouble get much more attention, esteem and support. So the project manager creates bad news, plays "the rescuer" and then is promoted to another job before the project is closed.

Systematic work and discipline alone do not help. A methodology, the observation of some basic laws of project management and a tool that fits for the project team are needed.

Henry Gantt (1861 – 1919) developed a methodology and its graphical presentation in the years 1910-1915. Military used this methodology during World War I. After the 1980s with the wide spread of PC's in software development teams, Gantt project management software became a state of the art. In the 1990s, the first groupware solutions were installed in large organizations. The Gantt methodology uses a timeline in an x-y-coordinate system. Bars for each work package indicate when the work is planned to begin, how much time is calculated and when it should be finished. The resources and the degree of completeness can be added to the bars. Bars indicate what work packages can be done in parallel. Important is what work packages require other activities to be finished first before they can be started. So the dependencies define a critical path with a minimum of time necessary for the project. This allows management to decide how helpful the desired result of the project will be at the time of completion. Eventually, the objective and content of the project can be changed. Staffing and resources may play a smaller role as 'a thousand people cannot build a house in one day'.

Gantt diagrams become difficult to read for larger projects with several project leaders. Some companies added the documentation of what person contributed how much to the project. This may be helpful. It may also make the Gantt diagrams confusing. An additional accounting and controlling tool might be a better solution. This is one reason why SAP integrated project management in its software.

After the year 2000, groupware solutions with full documentation and different views on the projects became possible and were widely used. Management and marketing can see the milestones, the accountant can see the budget and when it will be used, controlling can check the cost accumulated and the value produced, delivery can prepare its teams for bringing the product to the customers.

In 1958, Ben Tregoe and Chuck Kepner started a partnership and decide to teach others their method for problem solving and decision making.[88] Their company is now working worldwide. The company's methods are used by large companies and made them successful. Kepner-Tregoe is taken here as an example, may be because the author was trained so. The Internet site of the company publishes the major principles and the timeline.

When he author became an officer in a ministry in 1988, he found that governments use similar methods but often less systematic and with less emphasis. The idea of open government will hopefully give the public more insight in problem solving and decision making techniques of public administrations. Generally accepted standards have the potential to reduce the costs and increase transparency of public administration.

Projects are key activities for transformations in companies, administrations and the society. This book is not the place to describe in detail the integrated approach that transformation management and project management need.

Why is project management so important?

- Today, efficient companies and governments organize as much work as possible in the form of projects. All projects should be of short duration.
- In today's flat hierarchies, project managers need a broad understanding of what they do. Formerly big projects with hierarchical structures have changed to interconnected projects for special tasks. The project managers cooperate in a coordinated and integrated way for the benefit of the customers. The project managers share creative ideas of their projects with the other project managers. This might change content and milestones of activities. As the principle still holds that 'moving targets' are not allowed in a project a formal re-planning and re-definition of projects might be necessary.
- Project managers are motivators for their team. They give good examples for work efficiency. The objective is to begin and end meetings on schedule, to add quality assurance activities like a formal evaluation of every meeting at the end by all participants, and document all activities completely and in real-time. The 'culture' of a company is much defined by the working style of the project managers.
- Projects managers are specialists in their field and often qualified as certified engineers. Special institutes for project management were founded. In 2012, ISO-Standard 21500:2012 for projects was published. ISO reports on its web site "Project management is now big business. According to the Anderson Economic Group study commissioned by the Project Management Institute, over 24.4 million employees were participating in projects in 11 major economies in 2006. By 2016, this demand will exist to support 32.6 million employees in the same countries."[89]

Looking back the timeline, we see that many attitudes, methodologies and techniques invented in the 1950s and the two or three decades later changed the world. This is true for Gantt, for Kepner – Tregoe and many more. Today there are other people working to change the world with their ideas in the next 30 or 50 years. They are unnoticed, unless we search for them and support their ideas to become reality much faster.

.

The employees of IBM Germany gave a report of their experience in their book about the transformation of IBM Germany and the lessons learned (see box). They are convinced – to take only some ideas out of their story – that networking and integration are one key for success in the global information society. Another lesson learned is that planning and creativity should never be in conflict. No doubt, that experience is one motivation for agile project management. The persons needed for a company working daily on transformation should be enthusiasts of change. "Panta rhei" ("everything flows") was wisdom in Greece, China and many more cultures centuries before Christ (B.C.). Computing Science proved the validity of this ancient view with great emphasis.

## 6. HIGH PERFORMANCE SYSTEMS

May be another Sputnik shock will happen in 2013. China tested its Tianhe-2 or Milkyway-2 supercomputer with up to 3.1 million cores at nearly 31 petaflops. China developed its own front-end processor and its own software with special parallelism. The country works on its own chip technology.[90] It is reported that the new Chinese system will use 24 MW (megawatts) at peak when cooling is considered. 1 MW is calculated as US-$1 million per year. The goal for exascale is said to be 20-30 MW.

Mega = $10^6$
Giga = $10^9$
Tera = $10^{12}$
Peta = $10^{15}$
Exa = $10^{18}$
FLOPS = Floating Point Operations per Seconds;
Exascale System = 1.000 petaflops

This marks a new milestone in the research and development of fast computers. The infrastructure of supercomputers connected with fast networks positions a state in the global competition for the development of industry and research. The hunger for super computing power will not be satisfied as far as we can see today. The development of the batteries we need to drive future cars and to store the electricity we get from renewable energy need complex calculations of theoretical physics and chemistry. Enterprises need supercomputing for the competitive development of their products. They need commercial super computers with cloud computing capabilities working with optically controlled networks.

**IBM: Did you know?**
IBM earned 4,186 U.S. patents in 2008, becoming the first firm to collect more than 4,000 U.S. patents in a single year. IBM's 2008 patent issuances exceed those of Microsoft, HP, Oracle, Apple, EMC, Accenture and Google combined.
Sixty-five percent of IBM's women executives are working mothers.
All of the world's top 50 worldwide banks use the IBM mainframe. In the first quarter of 2009, IBM announced that its System z mainframe business grew 37 percent in emerging markets such as China and India.
(from: IBM Slides Mainframe 2009)

The competition to build the best supercomputer began in August 1963 when Control Data Corporation, founded in 1957 by a team of engineers among them William Norris and Seymour Cray[91], announced its 6600 supercomputer. As Thomas J. Watson recalls, this machine delivered three times the power of the IBM STRETCH machine and was cheaper. IBM cancelled its supercomputer program after some years and having been outnumbered by Control Data Corporation 15:1. [92]

The supercomputing centres provide a lot of information about the need for supercomputing, their configuration and the conditions for use. An important supercomputing centre is operated in Jülich for German and European users. (www.fz.juelich.de).

IBM is one leading supplier of these fast and big computers. In 1997, IBM's 'Deep Blue' computer calculated 200 million chess moves and won a man-machine competition with the world champion at the time Garry Kasparov. For the first time in history, a computer won such a match.

Fig. 19: IBM Processor for Scientific Supercomputing

In February 2011, IBM tested successfully its new workload optimized computer system Watson in a Jeopardy! contest with candidates Ken Jennings und Brad Rutter. The spectacular achievement is the ability to compete successfully with human experts. The hardware basis of Watson is IBM's POWER 7 technology interconnected in a massive parallel architecture. The software and the information retrieval out of a knowledge base with expert's information from a huge storage are as important as the hardware. IBM integrated information with different degrees of granularity to cope with the multiple levels of problems in the Jeopardy! contest. Possible correct solutions received a pervasive scoring with confidence values based on probability schemes.[93]

The supercomputing centre in Munich installed in June 2012 one of the fastest petaflops computing systems of the world (SuperMUC). With more than 155.000 cores and a peak performance of 3 petaflops, the system is available for all European researchers. A special migration system SuperMIG allows users a fast migration to SuperMUC. As the use of energy became more and more an

.

issue, SuperMUC uses a new solution: "SuperMUC uses a new, revolutionary form of warm water cooling developed by IBM. Active components like processors and memory are directly cooled with water that can have an inlet temperature of up to 40 degrees Celsius. The 'High Temperature Liquid Cooling' together with innovative system software promises to cut the energy consumption of the system. In addition, all LRZ buildings will be heated re-using this energy."[94] For more information see www.lrz.de.

In the late 1990s, ICT specialists began to "downsize" applications from big commercial computers ('mainframes') to servers, workstations or PC. Sometimes, radical cost cuts were possible. Early protagonists said, the mainframe will no longer be used. This prediction was not correct. Mainframes are improving, and some see a mainframe like a highly integrated cluster of fast servers. A mainframe is more as it allows to realize 'zero' failure data centre operations. In the 1970s, a failure rate of one hardware failure per month was proudly presented as an extraordinary good mean time between failures. Today's mainframe's hardware architecture allows a mean time between failure of years and decades. This was achieved by back-up of all components of the hardware and an additional warning system if hardware failure is possible. Simple clustering of servers with open source software does not allow such high quality.

Data centres with mainframes use system administration tools of premium quality. One such tool is Tivoli©. These tools do not only check permanently the mainframe and its components, the operating system and the sub-systems (like transaction monitor and database system) but allow administration of storage area networks, decentralized servers, Internet firewalls and demilitarized zones, clients (inclusive mobile devices) and the licence management of the whole ICT infrastructure of a company. Flexible application use and customization with change and configuration management in a cloud computing environment is impossible without such administration tools.

In 2020 the fastest computer system seems to be "Fugaku", a computer developed by Fujitsu with ARM processing units. Fugaku calculates with 450 Peta-FLOPS, 2.8 times the speed of "Summit" built with IBM Power9 chips. Fugaku runs with single precision calculation with the speed of theoretically 1 Exa-FLOP. If this proves to be correct, it would be the first Exascale Computer.[95]

# 7. COMPUTERS BECOME A CONVENIENCE IN CLOUDS

Between 2000 and 2010, the global transformation was enormous although unnoticed by most people. The adjective "virtual" was used to nouns to say that a "virtual X" is like a real X but it does not possess materialistic realization. So for X = bank, X = retail, X = shop people thought, that no real world objects are connected with it. That was not quite true as Amazon and other companies proved. These huge companies owed at least huge data centres and e.g. logistics centres. The unused computing capacity was much more than what some big companies required.

IBM started in about 2000 with the concept of "on demand computing". The idea was that customers "buy" computing capacity from IBM whenever they need it while paying only for what they used. Another idea was to reduce costs through server virtualization and shared services. Soon people understood that especially the power consumption of data centres could be reduced. This motivation was enhanced by environmental activists to a concept of GreenIT. Of course, everybody responsible for high performance computing was happy with such a motivation and the cost cutting consequences.

Companies like Apple, Microsoft, HP, Amazon looked for further possibilities to

- provide new services like e.g. maintenance of the mobile devices. Many users had problems to update the operating system on their mobile device. Others had lost their mobile device and wanted to know where it is, or if it could be 'destroyed' in some way or the other in the hands of the new possessor.
- E-books bought especially for Amazon's Kindle should not get lost in case the mobile device fails, breaks or gets lost.
- New companies (e.g. offering digital photo albums and rendering) got an opportunity to offer their services out of a cloud. As billing depended on the computing power used, these young enterprises enjoyed low fix costs and could cover the variable costs with the revenue from their growing customer base.

.

- Selling Hardware and Software came to a point where further revenue was hardly possible. Especially the old fashioned PC and even notebooks were no longer wanted. New needs of the customers could generate new revenue for the providers.

The development outlined in the previous chapters especially towards personal self-programming with graphical programming interfaces became easy and functionally rich. So three developments came together:

- The users could customize or program whatever they needed themselves without help from professional programmers.
- The huge data centres could provide even in peak hours all computing and storage capacities needed and sell this computing power in an "on demand" pricing. These data centres could even provide a premium service level with back-up, flying restart and uninterrupted service around the clock all year long. Billing of computing power depended on the capacity that was used. Several different billing models were applied.
- The Internet provided all the data communication capacity needed to use these functions from a distance.

The U.S. National Institute of Standards (NIST) published in September 2011 the following characteristic of Cloud Computing and a Cloud Computing Reference Architecture[96]:

The characteristics mentioned are:

*"On-demand self-service. A consumer can unilaterally provision computing capabilities, such as server time and network storage, as needed automatically without requiring human interaction with each service provider.*
*Broad network access. Capabilities are available over the network and accessed through standard mechanisms that promote use by heterogeneous thin or thick client platforms (e.g., mobile phones, tablets, laptops, and workstations).*
*Resource pooling. The provider's computing resources are pooled to serve multiple consumers using a multi-tenant model, with different physical and virtual resources dynamically assigned and reassigned according to consumer demand. There is a sense of location independence in that the customer generally has no control or knowledge over the exact location of the provided resources but may be able to specify location at a higher level of abstraction (e.g., country, state, or datacenter). Examples of resources include storage, processing, memory, and network bandwidth.*

.

*Rapid elasticity. Capabilities can be elastically provisioned and released, in some cases automatically, to scale rapidly outward and inward commensurate with demand. To the consumer, the capabilities available for provisioning often appear to be unlimited and can be appropriated in any quantity at any time.*
*Measured service. Cloud systems automatically control and optimize resource use by leveraging a metering capability1 at some level of abstraction appropriate to the type of service (e.g., storage, processing, bandwidth, and active user accounts). Resource usage can be monitored, controlled, and reported, providing transparency for both the provider and consumer of the utilized service."*[97]

In 2013 Microsoft offered Office 365 as a service in the Microsoft cloud. Further services became possible like the synchronization of documents through the Microsoft Cloud to all Windows based computing devices throughout the world.

Cloud computing is an up-to-date technology like object oriented programming or Web 2.0. It is no marketing idea lacking fundamental technological background.

## 8. ARTIFICIAL INTELLIGENCE AND ARTIFICIAL NEURAL NETWORKS

Artificial Intelligence (AI) and Artificial Neural Networks became an issue with much commercial possibilities. Depending on the commercial interests, speakers and companies define "intelligence" in their own way. This reaches from infinite possibilities to apocalyptic prophecies.

In 1950 Alan Turing published "Computing Machinery and Intelligence", proposing a Turing test to answer his simple question "Can machines think?". Software could be called "intelligent" if it is able to make a computer think in an indistinguishable way like a human. No computer passed the Turing test yet. In 1956 John McCarthy held the first academic conference with the notion "artificial intelligence". In the following years, expectations ran high, proposing breakthroughs regularly, but these expectations were never met. Of course, many alternative tests were proposed with levels of ambition depending on the personal commercial or scientific objective.

Pragmatic researchers worked on computers playing chess. This game was considered an example for intelligent reasoning. We now know that many simple chess programs use the "brute force" algorithm to impress the average chess player. Even IBM's Deep Blue computer only evaluated about 200 million positions a second and the idea was to beat sooner more chess pieces than the adversary. Together with a rough knowledge base of openings, this was generally enough to be successful. But it was no sign of intelligence. And an addicted human chess player values chess games with computer adversaries hardly fascinating.

Then the idea of "expert systems" was developed. These systems consisted of a rule data base defining the reasoning of the software, a knowledge data base for the ontology (in information science this is a structured, formal database with definitions, categories, properties etc. and their relationships) and backtracking algorithms, fuzzy logic and further inference concepts. The programming language was often LISP, without many possibilities of system integration. But especially the validation of the quality of the expert system lacked objective criteria. Their value depends mainly on the quality of the ontology data base. Especially medical applications gained much value with well-defined ontologies, a work for large international teams that required often many years.

Help for AI research came from the development of cheap and extremely fast computers, consistently structured storage (e.g. HANA) based on a scientific taxonomy, powerful search engines and graphical user interfaces. Between 1990 and 2010 many expert systems used this complex and powerful technology based on knowledge, deeply structured by human experts. AI became a breakthrough. Many of the systems developed at that time are of great value in everyday use till now.

Seen from the point of computer science, expert systems were rarely built with the concepts of AI as worked out some years before (like e.g. backtracking).

Progress in Artificial Neural Networks (ANN) accelerated AI to its immense power as we see it today. Researchers studied biological neural networks and developed a model of these with computer science notation. Tests proved that a simple processing model mapping only 60% or so of the biological facts to computer models is highly productive. Especially automatic learning was possible. For the Go game contest, automatic learning without human interaction proved its unexpected possibilities. Speech recognition and other applications needed and still need human help for the deep learning processes. But this help was commercially well spent money.

In 2020 such AI solutions are state of the art.

Currently there is no indication that computers will ever be able to think like humans. But nevertheless computers (realized as apps or avatars) are in a thousand ways helpful for humans and we probably did not even see a glimpse of the overall possibilities yet. Unions see the danger, that simple jobs will vanish and AI robots might substitute them. This depends how we use this technology. AI can help every worker, no matter what his qualification is, to be successful like it helps e.g. an insurance expert to find the right solution for his customer using the expert system of the company. Latest regulations must be implemented in such worker support systems to avoid time consuming courses for the human user.

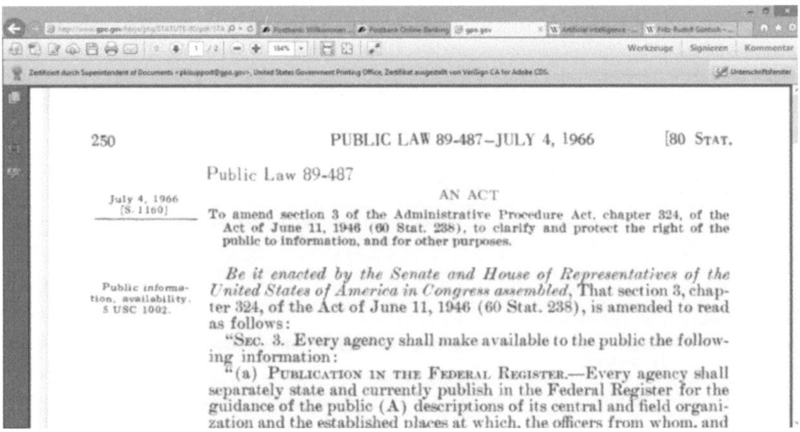

250 PUBLIC LAW 89-487—JULY 4, 1966 [80 STAT.

Public Law 89-487

July 4, 1966 [S. 1160]

AN ACT

To amend section 3 of the Administrative Procedure Act, chapter 324, of the Act of June 11, 1946 (60 Stat. 238), to clarify and protect the right of the public to information, and for other purposes.

Public information, availability. 5 USC 1002.

*Be it enacted by the Senate and House of Representatives of the United States of America in Congress assembled,* That section 3, chapter 324, of the Act of June 11, 1946 (60 Stat. 238), is amended to read as follows:

"SEC. 3. Every agency shall make available to the public the following information:

"(a) PUBLICATION IN THE FEDERAL REGISTER.—Every agency shall separately state and currently publish in the Federal Register for the guidance of the public (A) descriptions of its central and field organization and the established places at which, the officers from whom, and

Fig. 20: U.S. Freedom of Information Act, 1966

Privacy and freedom of information are basic human rights and existed long before the era of digital computing. Some enthusiasts of Chinese criminal stories may recall the illustrative activities of Judge Di. Even those readers, who are not friends of legal reading, will enjoy a look into this ancient world of administration. Judge Di's court sessions were public like those of many other cultures. Neglecting the aspect of public education and demonstration of power, the intention was freedom of information through the right to attend court meetings. Watching parliament debates and decision making, reading published budget plans was something our ancestors fought for when they created our modern democracies.

As an anecdote, we present a directive from the 19$^{th}$ century. In 1853, the Duke of Baden published a directive on the deletion of files in his paper registers. Probably privacy was not the reason. The Duke might have run out of space. All the same, the directive is interesting because of its detailed definition of the different duration times for storing documents that could still meet modern standards.

In 1966 the U.S. published its first Freedom of Information Act. This law could still be a model for countries that do not have comparable legislation in 2013. But: Many citizens want much more.

The political awareness of 'privacy' and "freedom of information' changed in time and degree of industrialization. When digital computing became a possibility, some saw the chances and other focused their attention on the challenges. As reported already, the chances were overrated. The challenge was likewise difficult to understand. Mathematical concepts, technical possibilities and legal ideas had to be brought together. At the U.S. National Computer Conference in 1974, Jerome Lobel from Honeywell Information Systems came to the conclusion that computer privacy is probably too expensive.[98] Today we know that the author was despondent, and the Duke of Baden had more advanced ideas.

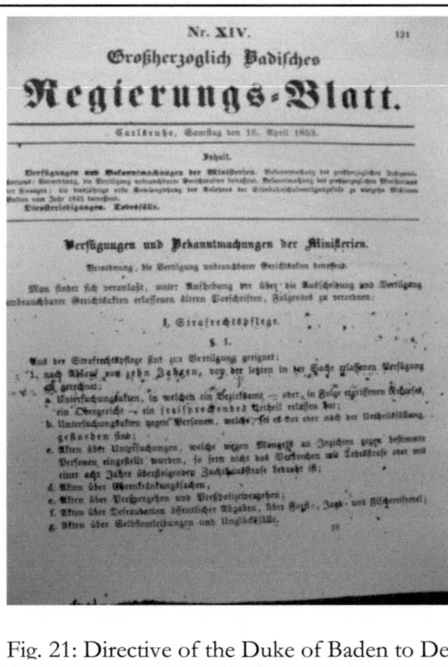

Nr. XIV. 131

Großherzoglich Badisches

Regierungs-Blatt.

Verfügungen und Bekanntmachungen der Ministerien.

Fig. 21: Directive of the Duke of Baden to Delete Files

The European Privacy Directives had an important cost cutting side effect. False and useless data are no longer stored and processed in private and public data bases. Carelessness and misunderstandings by processing personal data had sometimes dramatic consequences. The author recalls a case when sloppiness led to the insolvency of a company. Incorrect processing of personal data should never happen. Privacy regulation proved to be a benefit in several respects for the information society. The police forces e.g. learned to work with high quality data and so gained efficiency.

Like Spiros Simitis, pioneers of privacy regulation for the information and knowledge society searched in the early 1970s for a legal model to describe digital computing. The difficulty was to find a stable framework of definitions for the fast-changing software and hardware technologies and the associated organizational developments. The model found is of great value for a sustainable solution of the privacy challenges and for other areas like out-sourcing and cloud computing.

Detailed historic information is found in publications about privacy.[99] [100]

This report concentrates in several respects on the development in Europe. There the basic notions and legal concepts were found in the 1970s and were built on two important pillars. The first pillar consists of the committees and commissioners that were founded. The legal concept is the second pillar.

## DATA PRIVACY COMMITTEES AND COMMISSIONERS

This first pillar was introduced especially in Germany and France in analogy to the "Ombudsman" because the privacy laws had not only a legal but also a political dimension. It was a general view that a law forcing the citizens to go to the courts to claim their right of privacy was not adequate to digital computing. The public could not understand digital computing and its legal aspects. The courts would have needed a lot of expensive expert's opinions and testimonials to find out the status of data processing of the individual case.

In Germany, the law introduced state and federal data privacy commissioners with the right and the duty to check by their own judgment all state, federal or local government authorities without having initial suspicions or other concrete motivation. In France, the law introduced a national committee for information and freedom. Some laws gave the commissioners and the committees the right to veto projects. Commissioners and committees were quite helpful. To give an example, they found out that personal data could not be deleted in some police data bases, and that money was wasted as organizational procedures and safety regulations were missing. The commissioners and committees presented regularly reports on their activities and findings to government and parliaments. As these reports were published, data privacy became regularly (e.g. once a year) a topic of political discussions. Although this produced often changes of the privacy laws, the general concept sustained.

In 2012, the EU Commission proposed a unified privacy directive for Europe to avoid the negative effects of the differences of the regional privacy laws. As an example, the restrictions on the processing of social, medical, religious and criminal data developed in different directions. Even opposing data privacy regulations in the European Union hold in 2013. Italian law requests medical certificates to be written on a smart card and German law forbids just that.

These smaller differences totalled over the years to an obstacle to unified European Cloud Computing, E-Government and E-Commerce solutions. The European Privacy Directive requires for multinational and multiregional ICT projects to meet all privacy regulations of all regions concerned. In the light of the Single European Market, the proposal of the European Commission seems to be necessary in order to compete with the big U.S. cloud computing providers working often under U.S. law without exhaustive privacy regulation.

As the European Union publishes all legal development and discussions on its Internet site www.europa.eu, the reader is able to access up-to-date information.

## DATA PRIVACY REGULATION

The core of all privacy regulations is:

- Full Transparency: Every citizen should be able to know who processes his personal data.
- Binding on Purpose: Data collection and data processing is limited to a specific purpose for which the data was requested from the citizen or customer. The use of personal data for another purpose needs an appropriate legal basis (consent, contract, public task, vital interest, legal basis). Personal data which is no longer necessary for the initial purpose should be erased.

- Correctness and Fairness: Data processing is only allowed if it is fair, accurate and transparent.

"United Nations Resolution 45/95 of 14 December 1990, entitled "Guidelines for the Regulation of Computerized Personal Data Files," recommends that each nation develops regulations that require fairness, accuracy and transparency of data files, as well as a ban on using collected data for discriminatory purposes. Guidelines of the organization for Economic Cooperation and Development (OECD) adopted at the 523rd meeting on 23 September 1980, on the "Protection of Privacy and Transborder Flows of Personal Data" advise that any personal data gathered should be relevant to the use of that data and should only be disclosed to others in relation to those purposes. These guidelines also include a right of individuals to obtain and challenge any personal data that is held by other parties."[101]

The first pillar, the data privacy legislation, is discussed here only on the basis of European Directives to reduce complexity. Basic privacy regulation is laid down in the directive 95/46/EC from October 1998. Directives of the European Union are addressed to the Member States (MS). MS are required to change national law in conformity with the directive. Further special directives became necessary according to the progress made with the development of the Single European Market. When the telecommunication sector was regulated to allow e.g. privatization and pan European commercial activities and later mobile services, a special directive became necessary. [102]

The fathers of the EU privacy legislation unanimously were convinced that transparency and correctness of data processing is most important for citizens and customers. As today's means to gather personal data are not transparent to people without detailed knowledge of computing processes, privacy regulation begins with the collection and recording of personal data. After the collection, personal data may be stored, changed and given to third parties.

Data collected may be correct or incorrect. The question is how a person can find out what personal data are stored and distributed in connection with the person's name. This is especially difficult if the person's name is not directly but indirectly used through the household address, the number of the drivers licence, the iris pattern or the number of the social security card.

Article 1 of directive 95/46/EC defines:

.

*"'personal data' shall mean any information relating to an identified or identifiable natural person ('data subject'); an identifiable person is one who can be identified, directly or indirectly, in particular by reference to an identification number or to one or more factors specific to his physical, physiological, mental, economic, cultural or social identity".*

As for the processing phases, Article 1 of the same directive defines:
*"'processing of personal data' ('processing') shall mean any operation or set of operations which is performed upon personal data, whether or not by automatic means, such as collection, recording, organization, storage, adaptation or alteration, retrieval, consultation, use, disclosure by transmission, dissemination or otherwise making available, alignment or combination, blocking, erasure or destruction".*

These definitions of the general privacy regulation provided a stable set of formulations for all privacy regulations. On this basis, further legal concepts could be worked out. EU directives exist for telecommunication, e-commerce and e-government. Anonymous data is not protected by privacy regulation.

Built on these definitions, laws and private contracts define the circumstances when the processing of personal data is allowed.

Special regulations were found for national and public security.

Citizen shall understand whenever citizens' data are collected. Here, recent privacy regulations take into account that citizen understand more and more about the functioning of computers, e.g. through the use of their mobile phones.

Personal data may be processed if

- a law or another regulation based on a law allows it; special restrictions may apply to the duration of storage, the permissibility of distribution, the processing phases mentioned in the definition above;
- the citizen or customer voluntarily and with full understanding of the consequences agreed to the processing (of all or some phases) with full consent;
- a contract or a pre-contract situation (pre-contract: e.g. in the case of job appliance) defines the details of the processing of the personal data.

The main objective never alters: Transparency and correctness are to be maintained. Each citizen is entitled to understand the processing of his personal data. If a person agreed to the processing of persnal data, this person can withdraw this consent anytime. Of course, the withdrawal may have negative consequences like the discontinuation of service provisions.

Based on this framework, a set of political objectives arouse. Examples:

- If ICT applications can be developed and operated in a way that no personal data processing is necessary, the collection of personal data is not allowed.
- The German constitutional court ruled that the state is not allowed to spy in personal computers of individuals without a legal basis. The personal computer is seen as a part of the personal environment which is like a person's home protected by the constitution.
- Applications, that transmit personal data between the personal client computer and the server of a private company or of an authority, must offer encrypted transmission. The transmission may be generally encrypted.

The fathers of the data processing regulations saw from the beginning that companies and authorities do not always use their own computer for the processing of the personal data. Often, a data centre of a service provider is used offering greater efficiency and lower costs. The responsibilities of the three parties, acting together in such a case, are defined in the directive. For this purpose, the privacy directive defines the 'controller' and the 'processor'. The controller is the agency or the company the citizen interacts with to get his business and intentions cared for. The processor is the ICT service provider. So, the processor operates the computers for the controller. The privacy regulation shares the responsibility of processor and regulator in the following way:

- The controller is responsible for the content of the processing of the personal data. What data are processed, how long these data are stored, to what organizations these data are transmitted lies within the full and undivided responsibility of the controller. The citizen is not legally linked to the processor.
- The processor gets clear, complete and written instructions from the controller about how the personal data are to be processed. The processor has no right whatsoever altering the processing phases of the personal data entrusted to him by the controller.
- The processor's duty is to advise the controller about improvements concerning data security or data processing efficiency. If the processor finds out that the instructions of the controller are incomplete or false, the processor must request further instructions.

- Privacy law implements special regulations in case of criminal activities, police investigations, security services and public prosecution.

Now, the whole model of the regulation is visible. This model is a good basis for contracting outsourcing, privatization of ICT and cloud computing.

Today, privacy legislation faces some challenges. People give a lot of data away when they use to shop at only a few online shops, use one social network and one cell phone or smart phone. People usually use a credit or debit card and a bank account. They leave a lot of traces on their virtual trails. Only in order to understand the amount of data generated by one person, the reader can imagine that none of the data is deleted, and all data is put in one data base. Connecting personal data is quite successful by using only 'name' and 'day / month' of the birthday (or 'name' and 'year of birth'). A huge amount of data is available after some years. Cell phone technology registers all cells in which a person spent time as well as the duration of staying there. Shopping habits, favoured hotels, popular reading, travels made and financial adventures undergone are all registered in some electronic storage. Social networks and search engines of Internet providers add more data. There is currently no automatic connection between these data bases. This scenario is only provided to understand the potential.

Big data analysis methods allow digging through any amount of data within seconds no matter what the digger is looking for.

The European Commission tries to create a homogenous privacy regulation within all member states of Europe. The Commission intends to give special attention to threats of social networks and mobile phones. Politics promised the citizens that an electronic rubber will be provided to delete traces left without intention or imprudence in the Internet. Currently there is no technical concept for a general electronic 'Internet Rubber'.

The Commissions initiative supports the political objective to promote European commerce and create more jobs. As a consequence, member states would lose some autonomy.

Readers who would like to read a fiction about the possibilities of avoiding electronic traces could try "The Traveler"[103].

## CHALLENGES TO DATA PRIVACY

These privacy concepts are challenged in several respects.

The EU developed privacy regulation of outstanding rigidity. Nevertheless the privacy of European citizens is not overall protected. In the course of the discussions resulting from the disclosure of confidential documents by Edward Snowden, a public discussion arose showing that European regulation is of no practical importance. All important infrastructure in the EU, like e.g. mail relay servers and the technological basis for the use of the Internet, is obviously operated under the laws of the U.S.A. and / or by U.S. firms. So, the privacy of American citizens based on some rules of the Patriot Act may be better protected than that of European citizens although the U.S.A. does not publish a general privacy law like the EU.

To enjoy privacy and autonomy, the EU should build up Information and Communication Technology (ICT) companies and ICT infrastructure of its own. The market share of e.g. Apple, Microsoft, Amazon, Ebay and Google shows the difficulties that European competitors would face if they tried that build up.

Talking here about the history of computer science, we note the errors made by European industry between 1975 and 2000. Siemens, Nixdorf, Compagnie des Machines Bull, Software AG, Nokia and many more had a realistic chance to successfully rival U.S. and Asian companies. Lacking phantasy and being pessimistic, often even negative, about ICT, the leaders of the ICT companies followed the overall public awareness and the conservative political mainstream in Europe. Industry pearls were sold, sometimes on the foolish assumption that ICT is still strong in the EU as companies like IBM operated important affiliates in the EU. Many ICT jobs were lost, hopefully not forever. However, in 2014 European public is still not aware of the necessity of building and maintaining advanced ICT infrastructure and software technology in the EU. This changed in 2020 when it was necessary to become productive within the COVID-19 lockdowns.

Another challenge to privacy is big data. From a European perspective this technology is considered an attack to privacy. Europe risks losing another chance to become, what is written in the Maastricht treaty, the most advanced economic market of the world. Why is that so? We now learn that personal data collected in enormous quantities gives us new insight about "what" happens. Personal health data, collected e.g. by hospitals over many years, allow new research approaches to diseases we have not yet a remedy to cure and thus answering "why" things happen after having found out "what" happened. Currently, the EU provides no legal basis for such big data applications except personal consent of all patients or the conversion of all personalized data records to anonymous data.

"One organization, the American Society of Clinical Oncology (ASCO), aims to have ready by 2015 the production version of its CancerLinQ health system of a 170,000-record prototype unveiled in 2013. At a White House event in November 2013, Thomas Kalil, deputy director for technology and innovation at the White House Office of Science and Technology Policy, identified CancerLinQ as one of a number of healthcare systems that would aggregate large quantities of data to improve treatment."[104]

SAP and the SANA technology would allow the build-up of data bases with such data on a technical level. But it is quite unclear if the necessary legal and political support will be provided In Europe in time, too.

The European Program for Research and Development should mirror this support through substantial budget increases and new approaches in the spending of these budgets. The EU mostly supports feasibility studies whereas the U.S.A. government does not hesitate to pay for the development of new ICT solutions for pilot use or even productive use.

We now enter a new era of identity management.[105] The identity stored in the Internet about a person becomes important for all citizen, e.g. when looking for a job or running for a political mandate. The possibility exists that identities will be manipulated or forged. On the other hand, virtual identities may omit negative aspects of a person. The driving forces are the users of mostly mobile devices. They understand that working and living in a globalized economy necessitates the use of social networks, virtual institutions and worldwide shopping trading and teaching. Retrieving information from several content suppliers these users seek content syndication from content sources with personal credibility. Tweets are not just accepted from anyone but especially from persons with a background of trustworthy friends. Of course, a Facebook entry with many friends could be faked, so suspicion is always justified. Information could be forged. In any case the European privacy laws cannot cope with this upcoming new virtual reality.

A third challenge for data privacy is the necessity for security agencies to collect and process data. Computers did penetrate now in all areas of our life. The recipe of the 1970s till 2000 to let security agencies collect only the data listed in a security law, provides no longer an efficient possibility to secure public life. The simple solution taken by some security agencies to collect simply all data is at least in some countries not accepted out of economic and legal reasons. A practical concept that limits data collection to a level unanimously accepted by a vast majority was not yet presented.

## DATA SECURITY

Computer security measures always depend on the goods to be protected. Anonymous open data need another protection mechanism than the protection of general personal data, medical personal data, or national security related personal data. Data security became a science of its own. It splits today into different complicated areas of security like process isolation, network security, encryption and steganography, electronic or digital signature and smart card security. All these areas cannot be discussed in this history of computing sciences.

As for the values to be protected, general data privacy regulations are an easy to apply legal system. The basic ideas are the followings:
(i) For every ICT application, the values to be protected are defined by specific law. The primary purpose of data security for an ICT application is to specify and provide technical measures that fit together with well-chosen organizational measures to establish a nearly 100% secured application of such law.
(ii) Privacy regulation specifies that back-up is needed. Personal authentication of the personnel of the controller is "a must" in any case. Data processing steps are to be registered in a detailed protocol, and that this protocol's purpose is limited to state attorneys' investigations.
(iii) Data processing isolation means that data processing for purpose A must be totally isolated from data processing for purpose B. The same is true for data processing of different controllers.

(iv) Generally, there is a discussion what level of security is necessary. Usually, the general data privacy regulation requires "state-of-the-art" security whereas security for nuclear power reactors requires "state-of-research" security in some countries. This means that only security methods and models which are industry standards are necessary to secure general personal data.

**Design of Data Security for ICT Applications (EU legislation 2013):**

- Does the application require the processing of personal data? If no, it is obliged to work without personal data.
- What groups of persons are concerned? Define a security concept for each group.
- What is required by law or contract? Highlight these basic requirements.
- What is necessary to ensure complete security (state of the art)? Add what is necessary.

Many computer scientists believe that computer security is mostly a technical issue. Other security experts, e.g. former crime detectives or human resource managers, start with a thoroughly different approach. They see the risks in the human shortcomings like greed. The consequences are that security standards include technical, organizational, human, regulatory, and many more security relevant aspects.

1973 the American Federation of Information Processing presented the first checklist to improve computer security.

1977 Kenneth Wong proposed an onion shaped security model. The computer system in the centre is protected by several layers of independent security measures.

1986 the Federal Republic of Germany created the "Bundesamt für Sicherheit in der Informationstechnik (BSI)" by adding the responsibility for Computer Security to the former center for cyphering systems "Zentralstelle für Chiffrierwesen". Other nations integrated the responsibility for computer security in their national security agencies.

Since about 2000, information systems are a critical infrastructure for all governments and enterprises. As a principle, security strategists request the top management of any organization to decide on the risks that the agencies or the enterprises ICT systems could accept. Security experts recommend mathematical estimation models. One example is the following: Derived from non-ICT based security strategies (e.g. Kepner-Tregoes's method), computer scientists disassemble ICT processes in easy to understand and easy to secure units. For each unit the risk R is calculated as the product of the probability p of a successful attack and the maximum possible harm h. The sum of the risks

.

R=p*h is to be minimized. A chart with all units and the associated risk is an overall risk profile of the computer system or the ICT application. Management can then decide what risk level might be acceptable for the enterprise or the agency.

Enterprises and governments established Computer Emergency Response Teams (CERT) within their data centres. CERTs collect data. CERTs check the firewalls they use against attacks in the "demilitarized zones" of their network access computers to the Internet as well as the updates of the software providers for quality and consistency. Whenever serious attacks are discovered, each CERT sends an urgent report to all CERTs working together in one loosely coupled security group. This urgent report is not meant to be translated immediately in action as each CERT is responsible for the security of its data centres without restriction. The report is only to create attention and give early warning information. Of course, security agencies use their own way to cooperate.

Since about 2010, Cyber "war" became an important issue for governments and enterprises although no war is formally declared. Various attacks of malware to industry, agencies and companies were reported in the newspapers (e.g. Stuxnet virus). Criminals boast that they can hack a private e-mail address for US-$30 to US-$50. "Cybercrime services allow businesses (for example, street gangs with soldiers on the ground) to buy a supply line of stolen credit card data or bank credentials belonging to individuals or companies local to their area."[106]

Simple strategies and checklists provide no longer the security needed. A total isolation of ICT systems with databases of inventions, technological innovations, and sensitive customer data from the Internet is an efficient although rather un-popular protection. The importance of security awareness and training for employees is still growing. Companies understand that they must care for the acceptance of their security regulations. Many employees and high level managers in enterprises and public administrations do not understand restrictions. They expect that they can bring their own smartphone to the company (Bring your own device (BYOD)) and process sensitive company data and private data on the same device without isolation from the Internet and its social networks. BYOD opens the smartphone to many cyber-attacks. Security experts choose sandbox solutions programmed as Apps in order to maintain some basic security.

One of the main challenges for computer security and privacy seems to be that employees and even security experts don't understand any longer the different "virtual worlds" they are living in parallel. To give an example, employees of the public administration release information to a social network without official authorization. Sensitive political information is released on wikileaks platforms. Security experts may enforce one security measure like encryption as the major pillar for the security concept of their organization. To assess the quality of the encryption is impossible, as no measurement is provided by mathematics. Programming errors or other deficits in the encryption program may open doors to sensitive data that the security experts believed to be closed.

The revelations in summer 2013 of whistleblower Edward Snowden about the data collection and data analysis tools of the U.S. National Security Agency (NSA) resulted in a basic re-design of computer security. Firstly, the computer community understood that encryption methods and programs were vulnerable through build in backdoors by interested security agencies. The importance of open software concepts and open software distributions with a knowledgable computer community to check these was understood. Although this does not eliminate the basic problem of the inexistence of metrics to measure the strength of encryption concepts.

Secondly, software manufacturers became aware that their software could not be entered if and only if (iff) they designed and programmed it in an approbriate way. So home banking software in Europe was safe according to the EU regulation, iff no external and /or unchecked software modules were used.

## APP - SECURITY

Since about 2010, Apps are popular applications for modern operating systems. Apps are primarily programmed for touch screen user interfaces of smart phones and other mobile devices. These Apps create security risks as the programming languages for Apps are nearly as powerful as those used for servers. In contrast to servers, smart phones often do not possess program isolation. If no special effort is taken, Apps can access the directories, mails and application data on the mobile device. Apps can open program to program communication to any destination via the Internet as well. Data can be

.

transferred without authorization and without knowledge of the owner of the smart phone.

Of course, secure Apps are possible and available on the market. So the user is advised to check the security functions of the Apps and of the smartphones. The provision of safety certificates for apps would be helpful for the consumers. Neither the consumers nor the open software community are able to check the interiors of apps as thousands of them are on the market and new versions are published in rapid sequences.

## CLOUD COMPUTING

Many data privacy commissioners in Europe considered Cloud Computing from the beginning of the market penetration (NIST standardization in 2011) as a marketing idea and a threat to privacy. They think that personal data stored and processed in a cloud cannot be located and checked. Their concern is that personal data kept in a cloud are wandering around the world from one data centre to another thus becoming subject to many different privacy regulations and leaving traces in many places[107]. And they see the practically infinite capacity of a cloud as a dangerous tool to collect boundless information about everyone of us. Boundless information in an undefined data store sounds like uncontrollable means of evil data collectors preparing an undemocratic world. This threat is real as long as no legal obstacles ae established. In the European Union such regulation came into effect on May 25 in 2018 with the "General Data Protection Regulation (GDPR)"[1]. Before discussing details of GDPR, the discussion of clouds is continued.

If we do not consider careless contracting, the risks mentioned about clouds are not probable and the chances of cloud computing for better privacy and security are remarkable:

- Nomadic data wandering from one data centre to another is not part of the definition of cloud computing. Details are explained in the text above.

[1] In German: Datenschutzgrundverordnung (DSGVO). In French: Règlement Générale sur la Protection des Données (RGPD)

- Like any other ICT technology, cloud computing uses only well-defined computing processes.
- Cloud computing requires an extremely high quality of the the operating systems and the application programs. In a cloud such software serves numerous and heterogeneous clients of a cloud. In case of a software error, the information processing of all customers of the cloud would stop. That is why e.g. Microsoft developed an especially robust system (Azure) to meet these challenges. One patch level is necessary for all clients. Only one way of customization is possible. Before the cloud technology was introduced, these high-quality requirements were considered as impossible. Now we have it. The data privacy commissioners should welcome the fact, that cloud computing necessarily provide highest quality and best software documentation the marketplace ever offered.
- The documentation of cloud computing operations is usually also far better than much documentation of 'old fashioned' data centres. Otherwise the high service levels could hardly be delivered.

The European Commission understood the chances of cloud computing. In 2013, the EU Commission published a proposal to enhance European privacy regulation which currently demands operators of European wide software to implement the privacy regulation of the member states covered by their offer. The use of European cloud computing is reduced to specific application areas, a e.g. the privacy regulation of the European member states on health data is contradictory. Some member states allow the processing of medical data, some defined constraints that cannot be met by European wide cloud computing.

A cloud specific threat to computing security is a possible attack from one user of a cloud to other users of the same cloud.

The following examples show the use of public or hybrid cloud computing by public administrations in Germany in 2014 based on the current privacy regulations:

- Management of mobile devices.
- office Automation,
- telephony based on VoIP for small agencies,
- managing accidents, disasters or crises (e.g. aircraft accident, earthquake), disaster recovery,
- public information services like tourist information, press release in the Internet,

- Freedom of Information of Environmental Data. EU regulation gives all EU citizens a right to get environmental data for free.
- open data and open government portals,
- service-portals.

## GENERAL DATA PROTECTION REGULATION (GDPR)

The European Commission, the Data Protection Commissioners and the Member States of the European Union complained about the drawbacks of the European Data Protection directives:

- American Companies like Facebook and Apple did not mind the European Data Protection regulations. They continued collecting enormous amounts of data without giving their customers any insight in what was stored and for how long. The right of every citizen to have his personal data erased or rectified (corrected) was not provided.
- European and foreign organizations did not take the privacy regulations seriously. Penalties were modest or neglectable. Whenever data privacy measures might cost money, senior management decelerated the efforts of the responsible teams in their organization. This created frustration and disappointment on part of the people within the organizations and on part of their customers and the citizen.
- The Member States of the European Union applied a different interpretation of the data privacy regulations. Some allowed the organizations within their territory nearly limitless data processing, whereas e.g. France and Germany considered privacy an important and basic right of all citizen. A rigid unification of the regulation was necessary. Any kind of interpretation should be practically completed.
- New technology like Artificial Intelligence, Speech processing, Picture Processing, Quantum Computing had to be included into the regulations. The specific processing algorithms connected to these applications were not included in the old regulation and had to be integrated, too. The historically used concept of "Data Sets" could no longer be used to define the limits of Data Privacy. Most historic

concepts were fragmentary like e.g. the right of objecting to data processing.

In 2014 to 2018 much important work was done to solve these problems. Finally, a consistent regulation, the GDPR, was accepted by the Member States, the European Commission and the European Parliament. It was published as EU Directive 2016(679); all EU legislation is available on the Europe server europa.eu. Employer and industry associations criticized the GDPR especially because of its remarkably high penalties. But that proved to be a success factor for the GDPR. Finally, data privacy was considered important.

## FUTURE OF PRIVACY REGULATION

In 2020 many organizations still complain about high costs and inadequate over-regulation of data privacy in the European Union. Such weeping seems to be not successful. In the contrary. Parallel to an ongoing consolidation of the GDPR, new needs for further regulation are discussed.

- Applications based on Artificial Neural Networks (ANN) and Artificial Intelligence (AI) analyse personal data with speed and accuracy that human experts cannot compete. Diagnoses e.g. based on X-Ray pictures cannot be evaluated by medical personal in parallel to the automatic analysis because they lack time and expertise. If the application based on AI and ANN is not permanently maintained and updated, the result of the automatic diagnoses might be wrong. And nobody would find out these maintenance errors, as no practitioner is able to understand the software nor the medical facts. Thus the request arises (1) to constantly update all software, (2) to document the software in a transparent and easy to understand way and (3) to allow all patients (in general: all persons concerned) an additional inspection by a human expert.
- Quantum computers work based on quantum mechanics with quantum formulas. Because digital computing is rather complex, quantum computing is a field with only a few experts. The request to

document all applications of quantum computers in a transparent way is self-evident.

- Speech recognition can be used by all users of Apple's Siri, Google's Android, and Microsoft's Cortana. The process of such speech recognition is a stepwise perfection. On the user's screen, the user can easily watch the step by step processing. The above-mentioned requests are open for everybody's evaluation.

  Speech recognition demonstrates the process of artificial learning (deep learning) based on billions of personal user data. The software first produces a voice to text translation, then it improves the words recognized according to the personal use of such words in the past and possibly based on the data on the smartphone. Then the whole sentence is analysed and plural, verb use, etc. is improved. Thus e.g. the use of dialect is possible during dictation. The result is text with best English, German, French etc. Of course, analysis of dictation and writing behaviours of people allows the software of the Tech Giants deep insight into the psyche, intellect, attitudes and wishes of all customers on the background of their personal social network. This motivates the European Commission and the European Member States to create and pass legislation that allows the European citizen to own their own data. They will e.g. be able to carry their personal data from one social network to another. Erasure of stored data etc. should follow the rules of the GDPR.

- There is too much criminal activity done with the Internet. Identity theft, fake news, cybercrime, SPAM, etc. are not only 5% but often 50% of the Internet data flow. Children cannot behave like children normally do in the Internet. This chaos must stop.

# 10. ICT MANAGEMENT, OUTSOURCING AND PROCUREMENT

Since the beginning in the late 1960s, computers were a scarce and expensive resource. Companies shared their computer with others. Often public agencies were allowed to use private ICT equipment of local companies. The use of other people's computers was seen like the use of their car. Whenever a team used a computer in the 1960s, the new user came with his punched cards in boxes, switched the computer on and took printed paper, newly punched cards and everything else with him. Nothing was left in the memory of the computer.

## CIO ORGANIZATION

In the 1970s, computers had notable storage. Sometimes they were connected to a simple network and the idea to measure the use of a computer with a counter of the elapsed CPU time came into the mind of the manufacturer. When operating systems became functionally powerful, they were no longer for free. The disk files or data bases were well structured and of value for others. So costs had to be shared, data had to be secured, backup und archiving were issues and companies with a data centre needed more than a hand full of ICT experts.

In the 1980s and 1990s, the professional ICT in a company or administration became more and more complex. An ICT department was integrated under the direction of a Chief Information Officer (CIO). The department usually had a team for networking, one or more software development groups, a team for the supervision of the ICT applications, and the data centre's system specialists and operators.

The CIO was a member of the management board of some companies. Especially banks and insurance agencies used such an organization. In some companies, the CIO worked as a high-level manager in the hierarchy of the

Chief Financial Officer. Around the year 2000, the CIO became more and more powerful. At the same time 'out of the box' application software (like some modules from SAP and Microsoft SharePoint) became easier to customize. Microsoft released software that was easy to install and to operate with the objective of 'zero administration'. So many departments in companies and many agencies in the public administration built their own data processing infrastructure. After 2010, the departments saw cloud computing as a realistic alternative to in-house delivery by a CIO administration. The costs produced by the CIO administration came under intensive examination.

The justification of a CIO organisation was undermined by the general knowledge of nearly everyone using a home PC and a smartphone. Using object synchronization like in SkyDrive of Microsoft and programming one's own Internet site with the means of a graphical programming tool set, departments became autonomous. The CIO was not really needed any more. Therefore, some CIO was tempted to define short project cycles. Others gave the project management to the departments and restricted themselves to licence management, product management and in-house consultation.

In 2020 the CIO was much less important than in 2015.

Since the 1980s till today, the outsourcing of the development of ICT application software and the outsourcing of the operation of an ICT application cuts costs. The challenge of outsourcing is the build-up of a new infrastructure.

In the 1990s, network standards emerged with Virtual Private Networks (VPN). That favoured outsourcing of company networks either because the company wanted to migrate to a new and modern standard or because the company had migrated with its own experts and wanted to cut costs.

After 2000, a general outsourcing of ICT as a computing infrastructure began. The ways and means changed. The trend to more outsourcing will continue as cloud computing cuts even more the costs and at the same time improves the service level considerably. It may even reduce the company's fixed costs and allow more flexibility through variable costs. These are good reasons for a CEO to seriously consider outsourcing as an alternative to in-sourcing.

There are many ways and strategies to contract computing services. "Experience around the world shows that roughly two-thirds of outsourcing agreements fail to deliver the targeted benefits or create the hoped-for-value for shareholders."[108] The confusion seems to range from outsourcing seen as a way "to reduce costs by transferring portions of work to outside suppliers" to "buying cheap foreign parts rather than manufacturing them at home".[109]

History is of value. That can be seen in this context. None of the ICT pioneers promoted simple concepts like those followed obviously by many companies that outsource their ICT.

ICT is complex use of technology and must follow strategic plans of an organization. This requires that the management board defines objectives and installs people that can translate these objectives in sustainable ICT strategies.

It is no strategy to cut costs, as cost may be low already. In such a situation, cutting costs would injure a successfully working ICT and endanger the organization. Usually, ICT helps to cut costs in other areas. So cutting ICT cost is counter-productive.

The procedures of the public administration are more transparent than those of private companies. In the public administration, the following dangerous ICT strategy is often used. After finding out that the agency itself has no people to plan, build and run its ICT, some agencies try to formulate a call for tender. The basis of this call for tender is not an ICT strategy but a formal study of the procurement regulations.

The agency is tempted to write whatever is on the wish list into the specification of the call for tender. As procurement regulation requires the definition of different lots for services that are no necessarily part of an integrated package, agencies define several lots. Often they forget to look for a senior project manager to supervise and coordinate these lots, sometimes hoping (as there may be no budget available) they can coordinate the lots themselves. Typical lots are one for software development, one for the operation of the complete ICT application after its release from the responsible department and finally the maintenance of the application software. Procurement law requires that all lots are evaluated separately according to its cost-benefit characteristics. This procedure may have the following result:

- Lot 1 is given to a software team programming with open source modules.
- Lot 2 is given to a data centre specialized for the operation of Microsoft or SAP software.

- Lot 3 is given to another software team specialized in the programming of Oracle applications.

Obviously, this kind of outsourcing is not successful although all formal procurement regulations were followed. The interdependence of the three lots was not seen from the beginning, and an ICT strategy was not defined.

The author implemented several successful outsourcing strategies and will give an overview based on a scenario.

## SCENARIO: OUTSOURCING AN OFFICE SYSTEM

This scenario consists of the outsourcing of office automation systems of a state administration with n out of 10 ministries (n<10). Each ministry is supposed to be the head of a specialized administration like environment or justice. The task is to procure a unified automation system according to a concept (or design) and standards that all 10 ministries defined. The ministries that take not part in the outsourcing project should use the same standards. An easy and seamless communication is required. – The scenario is equivalent to the outsourcing of the office automation systems of a large group of companies (like a holding) with several rather autonomous group members.

**Checklist Outsourcing**

- Design the ICT of the organization concerned.
- Document all reasons for insourcing and outsourcing.
- Deduce an outsourcing concept and document the main characteristics (e.g. open data without privacy, open source or licences provided by contractor, entrepreneurial liberty for the service provider, budget frame, timeline, service level)
- What happens, if outsourcing fails? What characteristics does "failure" have?
- Search for appropriate legal framework.
- Define project and "Go!".

The first task to be completed is the consensual documentation of the concept and the definition of the standards that all should follow. The decision includes a documentation of the individual freedom of the specialized administrations as well as entrepreneurial freedom of the service provider.

Having this task completed, legal considerations start. The office automation package of company A is defined as ICT standard. To avoid criticism because of favouring company A and being discriminative towards all other companies, a rule is set in the standards that all other software package may be offered, if and only if the supplier guarantees a 100% compatibility for the next 5 years with the software package of company A. If this is impossible, the contractor should decide if the necessary licences will be provided to the service provider, or the service provider contracts these licences without the contractor's knowledge and as part of the prize per seat.

In this scenario, all ministries agreed that the budget used for the current in-house operated solution should not be changed. Especially, the budget should not rise.

The next major decision to be taken concerns the kind of procurement procedure to be used. With some simplification for shortness sake, the following possibilities exist:

- The solution may be discussed in a dialogue with several companies. The companies need to be searched and defined.
- The required solution is described in a final and irrevocable specification. Changes are not possible after the publication. The administration evaluates all bids according to a catalogue published with the specification.
- The solution is fairly and non-discriminatively found in a negotiation procedure. This negotiation procedure is only allowed if the call for tender contains aspects that cannot be defined and settled before the publication of the specification. The negotiation covers only the subjects with that well defined and well documented uncertainty within the specification.

A market evaluation will show that the costs of outsourcing depend much on the flexibility of the client to adapt to the computing infrastructure of the service provider. If this service provider uses an infrastructure compatible with the standards defined by the n ministries, cost saving may be possible, e.g. by rapid migration to the new service. If the bidder uses the same computing machinery like those in the in-house data centre, a better price may be negotiated.

These considerations lead to the application of the negotiation procedure.

All parties negotiating outsourcing contracts should be fair and transparent. A long time partnership will be established. Most probably, both parties will make errors during this partnership. If the negation climate was tense and the prizes tough, dramatic conflicts might result during the contracts lifetime.

For public procurement, transparency and non-discrimination are compulsory by law. The publication of a draft contract satisfies this requirement and provides further advantages. For private companies, this procedure is also recommended. Otherwise, bidders tend to explain their offer with buzz words in Power Point presentations without fully verbalized text. This procedure lays the foundation of endless discussions about what was meant.

If the purchaser supplies a draft contract, bidders gain lots of time to calculate risks and chances. Bidders can especially calculate

- the conditions on which they can provide their services
- the prizes of the different components (like project management, software licences, operation costs, reporting, technology refreshment after x years)
- the service level (like performance measured for Internet access, e-mail send and receive times, dialog quality, archiving standards, update cycles)
- help desk functionality and service level.

The call for tender should be precise on all these matters and define the absolute minimum and maximum of what is expected from the future service provider. If the bidder can offer more service and if this offer is appreciated, transparent negotiations are almost impossible. In that case, the negotiators would have to evaluate the offer that exceeds the limit. This is quite impossible without losing transparency and giving up equal treatment of all offers. After all, bidder and contractor can hardly calculate a budget in a situation with several unknown variables.

Under these conditions, the negotiation can be restricted to the prizes offered. After only a few meetings that are used to prize the special conditions of each bidder, the negotiation enters the final step of the "last and final offer".

After having chosen a service provider, the customers are often astonished to see other people of the service provider than the negotiators arrive in the meeting. This is a usual way to handle contracts in large organizations. The contract was handed over to the 'delivery group'. Customers with some experience foresee this during the negotiations. They include all promises in the minutes of the negotiations. These minutes become part of the contract although with low priority. Minutes should always be signed by both parties. Any time during the negotiations, the customers may request that the sales force contacts its delivery service.

Such procedure is only possible if the customer does not press for time. Experienced negotiators know that the party pressing for fast contracting is usually the party having the disadvantages on its side.

Delivery will then work out a migration path from the current office system provider to its own provision. Part of this migration path may be an assessment of the client's local area infrastructure. This will take time. If the purchaser established trust during the negotiations, the assessment may be limited to a few local area networks.

The customer is well advised to see all concepts fully documented before any computer systems are installed and used. Privacy regulation in the European Union requests this by law. After the documentation of the new system is accepted and possible tests were successful, the customer allows the service provider to install and operate the systems with real data instead of test data.

Migration from the old system to the new system should be contracted to happen seat by seat. If major problems arise, payment should be withheld. The "seat-by-seat" release allows an easy regulation to withheld money. As the bidder should calculate this in the prizes offered this migration and release procedure is part of the draft contract published with the call for tender. If no migration problems exist, the release of any number of seats is also possible with a simple declaration in the records.

Customer and service provider agree that the project managers of their companies are entitled to unlimited agreements. Both project managers coordinate their position within their organizations, of course prior to signing the agreements. This allows up-to-date and fast decisions. By the way, the party that drafts the minutes of the meetings gets an advantage. So, it is recommended that this is settled in the negotiation.

Procedures for a termination of the contract - after the time foreseen in the call for tender or after serious deficits of the services provided - play an integral part within every contract. These procedures are part of the draft contract in the call for tender.

Customers should be prepared that employees of the service providers may possibly change their jobs after two to three years whereas the employees of the customer hold their posts for a much longer time. The outsourcing project should plan these job rotations and establish a data base with all the information needed for the new team members.

Careful documentation and clear leadership by the customer is vital for a successful partnership. Many public agencies expect leadership from the service provider. They are often disappointed. Delivery departments of service providers do not necessarily check the market for better solutions. If leadership from the service provider is expected, regular information meetings could be planned in the contract. The service provider can be obliged to inform the customer about new technologies, new possibilities for data security, new functionality, and cost reduction possibilities. The final decision on the changes to be agreed upon is the customer's responsibility. Outsourcing does not work satisfactorily if the mutual responsibilities are not respected.

## USE OF OPEN SOURCE

Politicians see open source concepts as a way to create jobs. CEO's see it as a way to reduce costs. The open source community considers open source as a way to secure freedom and quality in the Internet. These opinions are not necessarily and generally true.

In 2014, many open source products are available in the market or in the Internet. Operating systems, office automation software, accounting software and workflow management software of high quality, and a lot more software are available. However, open source software is not bundled and managed in a uniform way. The difference to the product suites of companies like Microsoft, IBM, HP and others is that testing, integration, customizing and the administration of the operation are much more complex subjects than with proprietary software. "Zero administration" concepts are not part of the open source portfolio. No software maintenance is provided in an organized way.

Managers who want to use open software should see the total cost of ownership. The software licences, the team to maintain and operate the software and the technology refreshment define the costs of the ICT solution. Some open source products are updated constantly by 'the community'. All organizations using that software constantly check and evaluate the new versions in order to find bug fixes, security updates or simple functionality improvements or changes. This is cost intensive.

History shows that open source software is not maintained and may vanish from the market. It should also be noted that open source communities are not necessarily interested in seamless migration from one version to the next.

The open source community owns credit for discovering many errors and hidden functions in software of which the programming code was published in the Internet. The limits of the power of the community become evident when looking at the vast offers of apps with millions of lines of code in modern app stores. Examples showing the limits of the quality of community checked open source software are a) Pretty Good Privacy (PGP) and b) OpenSSL. PGP was initially developed and published as open source software to protect personal communication from security agencies. Nevertheless, it was for some time published in a version that did not encrypt at all due to software errors. As encryption is considered military equipment in some nations, the widespread distribution of PGP and comparable products is difficult. OpenSSL versions 1.01 to 1.01f were insecure for more than two years.[110]

Open source software is just one possibility to increase the quality of software. Commercial and proprietary software with defined liability, maintenance cycles and consequently appropriate prices became the most important pillar of today's software industry.

In 2014 the following software is of general importance in the area of open source software. Much open source software is distributed within limited communities like e.g. data centres (e.g. administration software for licence management) and public administations (e.g. encryption software, workflow and collaboration software e.g. for coordination of the many working groups of the European Administrations) and not mentioned below.

Examples for open source software:

- Open Office and Thunderbird as an open – but functionally limited - alternative to Outlook from Microsoft,
- MySQL data base,
- Apache and other modules for Internet services of data centers,
- Eclipse for software development,
- Operating Systems for simple phones were developed on the basis of UNIX versions. This accelerated the development and cut the costs for some companies. On the other hand, NOKIA lost its leading position in the cell phone market anyway. Furthermore, unlawful intrusion into the cell phones was simplified, because the technical platform of manufacturers was similar.

## PROCUREMENT

People procuring information and communication technology (ICT) need an in-depth understanding of the technology and the products offered in a competitive market place. Although procurement regulation formally binds only public administration, the principles of public procurement are often also followed by companies. The most important objectives of successful procurement procedure are

- to get value for money,
- to procure on a competitive market,
- being transparent to all tenderers and
- being non-discriminative.

Procurement means that one organization, the one calling for tender, and another organization, the tenderer for services and goods, agree on a contract. One difficulty is, that legally contracts can easily be settled. Such settlements need not be in writing and not terms and conditions may be detailed after having entered into the contract. Procuring complex products and services like ICT requires transparent procedures to avoid costly misunderstandings. Contracts should only be entered when both parties have clearly defined their respective obligations.

The objective "being non-discriminative" mentioned above requires a non-technical language in order to allow different competing companies to offer their goods and services. To do so is difficult as in the ICT market place providers often use product descriptions of their own (instead of the neutral language of computer science) with more or less explicit promises. This history of computer science explains the reasons for that. New products and services are often not fully developed when ICT providers offer them on the market place. Detailed development is carried out with the first users and contractors. To be successful in such a situation, sales people necessarily often describe visions instead of available services.

Negotiations either require a lot of time to write down all details or a lot of confidence and an understanding of the general goal common to all partners. Understanding technology is not much different from understanding people.

.

The development over the years should be observed carefully. This is what the history of computer science is about.

Joint development is a rare case. Usually, organizations that procure ICT need the services and products in a competitive quality (service level) and as fast as possible in order to achieve their non-ICT objectives. Careful and detailed negotiations are the usual way to procure complex and new ICT.

The art of negotiations is important in ICT procurement:

- Be fair and transparent. Negotiations need an atmosphere of confidentiality.
- Don't abuse errors of your partners in negotiations. Defining complex matters in joint negotiations will not be possible without errors of all participants.
- Leave the final decision open until the negotiations with all partners came to point where all offers can be compared in an objective way. For such comparisons numerous checklists and handbooks are available.[111]

The procedure of public procurement is as follows:

- The requirements and the need (demand) of the agencies is defined. This includes the accumulation of all goods and services connected with the requirements. To split requirements into several procurement procedures is best practice if the content of the requirements is not unreasonably divided into conflicting calls for tender and/or offers.
- The value of the procurement is estimated. An agency cannot procure iff the budget necessary is not available. Tenderer in a public procurement process can be sure tht the budget is available and the effort to place a successful bid is not in vain.
- The legal procedure is defined.
- All documents necessary for the definition of the procurement should be written. Then the procurement could be published and the documents distributed according to the legal procedure.
- The agency places the order to the successful bidder.

# 11. OPEN GOVERNMENT AND NEUTRALITY OF THE INTERNET

Around 2005 to 2010, politicians understood that the information and knowledge society is not easy to govern. The Internet allows citizens and enterprises to gather a lot of detailed information instantly. This information allows interested groups to formulate their interests professionally. Interest groups are even able to work out alternative proposals to politically proposed solutions e.g. for law making, the conversion of military used areas, the protection of environment, building new streets and highways and for tax policies.

Many politicians used the Internet to support their election campaign. A prominent example is Barak Obama. After his election in 2009, he supported the foundation of an open data portal[112] for the U.S. Australian and European governments soon followed. Even mayors found that bringing transparency to their plans gives them a chance to understand early what the voters think and what interests are at stake.

## OPEN DATA

Western democracies traditionally offer a lot of open data to the media and the citizens. Among these are all legal regulations and especially the budget, protocols of the parliaments, calls for tender and the outcomes, some decisions of the governments, a flood of statistics and rankings. International organizations use open data like e.g. the World Monetary Fund about Government Budgeting.

Open data strategists expect more. They want the open budget of governments, the opening of all data bases of the government that do not enclose secrets from citizens or enterprises, internal memos, concepts, studies, and freeware to versatile computation. Secrets are privacy protected personal data or business data of enterprises coming from sovereign activities (like the revenue service) of the governments into public administration's data bases. There are different political opinions about the sensitivity, the amount and the granularity of the open data to be released to the public. [113]

The European Commission's advisers expect that new businesses with a turnover of several billion Euro per year could be created with open data initiatives.[114] The details of the calculation are not known. If an analogue is drawn to the business created by GPS, it should be noted that GPS was built with enormous initial investments.

Scanning open data portals, the question arises what private person could analyse thousands of flat data files in order to work out an individual political opinion. Some think that this is quite impossible. One solution is seen in linked open data. Linked open data files automatically join with all flat data files that are stored somewhere in the open data portal whenever this makes sense.[115]

Information scientists stress the importance of the use of technical open data standards like "Comma Separated Variables (CSV)". List of such standards can be found in the Internet.

Groups like the Sunlight Foundation published criteria catalogues for open data solutions.[116] The Sunlight Foundation proposes also technical standards.

## OPEN GOVERNMENT

Open government allows citizens to voice their interests in many ways. Citizens sending e-mails to politicians or posting a message on a blog use classic Internet media. More elaborated Open government concepts allow an anonymous or personalized vote on a political or administrational agenda. Further concepts like Liquid Democracy are tested and products are available in the Internet[117].

Open government concepts are designed to test public opinion on political decisions like building a nuclear power plant, a new highway or a chemical production plant. Debates about housing concepts and the shape of new suburban places can easily be given to publics' opinion by open government portals.

More far reaching open government concepts need to be integrated with current democratic systems. The reason is that open government activities of governments put pressure on Members of Parliament. If a new law is proposed by the government, it should first be discussed with the Members of Parliament. Otherwise parliament is not entirely free to form an opinion. Furthermore, some political subjects are not popular. Early public debate on these subjects may not help to reach an appropriate decision.

Although Open Government portals may have registered users, there is no certainty that only those people voice their opinions that take part in the relevant political region or body. There are technical possibilities to provide anonymous participation e.g. by ticketing systems with virtual tickets being provided by local government. Such concepts need further political discussion.

## INTERNET NEUTRALITY

Net Neutrality is a historic characteristic of the Internet. It means that all Internet service providers transmit all data to all recipients at the same speed and the same costs. In the past, non-discriminatory treatment allowed the fast growth of Internet based services and of start-up companies. Google, Amazon, and Facebook might have been less successful if special charges or bandwidth restrictions would have been applied for new providers.

Net neutrality is the basis of the free use of the Internet. New ICT standards and platforms, new services, and new service providers could prosper in a free virtual environment. This free development of opinions with the use of the Internet could be challenged if some companies (like big social networks or Internet shops) get special rates and / or special transmission possibilities from telecom companies.

Net neutrality as a principle is in danger. The telecommunication companies do not get the money they need for investments in the Internet infrastructure. According to media reports, big Internet service providers support the telecommunication companies for years by the provision of powerful communication links. Public debate on net neutrality seems necessary to decide if the communications market is left in that respect without regulation.

Net neutrality and freedom of the Internet does not mean that the Internet becomes a place without legal regulation. Copyright law, licence regulations, criminal law and most other law is valid for transactions in the Internet's virtual world like in the real world.

In 2013, the German Telekom started an initiative to demand different prizes for the use of their telecommunication lines depending on the amount of communication traffic used and the partners contracted. Critics assessed this proposal as a risk for Internet neutrality. German Telekom wanted to exclude the use of Telekom's own services (called T-Entertain which is TV, Internet Flat Rate and a kind of Telephone Flat Rate) from the calculation of the more expensive communication traffic. In consequence, the European Commission discussed a European law to secure the Neutrality of the Network.

The political expectations connected with Open Data and Open Government take a free and neutral Internet for granted. Without becoming pathetic, access to the Internet seems to become the importance of a basic human right.[118] "Digital divide" is a threat to any society looking for more prosperity, progress and democracy. As far as economic development is concerned, liberal economic experts think alike.

To design a complete legal solution of these different interests and possible conflicts, the following legal activities might become a subject of political debate:

- privacy regulation
- freedom of information acts or government transparency laws with definitions of open government data
- Internet neutrality regulations.

For more information about the Internet and net neutrality, a book of Barbara Schewick[119] is recommended for further studies.

Net neutrality became a political issue in April 2014 in the U.S.A. and in Europe. According to the Internet Association for Assigned Names and Numbers (ICANN) the United States Government will refrain from access control concerning the ICANN root servers.[120] In Europe the European Parliament discussed a proposal of the European Commission on net neutrality. The EU Commission states that there were "no clear rules on net neutrality at EU level today, leaving 96% of Europeans without legal protection for their right to access the full open internet."[121] Critics say, the proposal of the EU Commission should be more precise. They think, the proposal will allow to elvels of Internet service, one fast and commercially sold at high prices, the other with low service level at lower cost. Thus start-up companies might have more difficulty challenging the bigger companies and non-commercial organizations might find it impossible to reach the citizens.

# 12. INDEX

## P



## Q



## R



## S

## 13. LITERATURE

---

[1] Readers are asked to report errors to the author by e-mail using the address given on the second page.
[2] Stephen G. Nash, History of Scientific Computing, ACM Press, New York, 1990
[3] R. W. Hamming, We Would Know What They Thought When They Did It, in: N. Metropolis et al, A History of Computing in the Twentieth Century, Academic Press, New York, 1980, page 8
[4] Paul E. Ceruzzi, A History of Modern Computing, 2nd edition, MIT Press, Cambridge Mass., 1998, 2003
[5] Georg Schäfer, Mythen, BoD Norderstedt, 2006
[6] Max Bense, Einführung in die informationstheoretische Ästhetik, Rowohlt Taschenbuch Verlag, 1969
[7] James W. Cortada, The Computer in the United States From Laboratory to Market, 1930 to 1960, M.E. Sharpe, Armonk, 1993
[8] N. Metropolis, J. Howlett, Gian-Carlo Rota, A History of Computing in the Twentieth Century – A collection of essays, Academic Press, New York, 1980
[9] Adele Goldberg, A History of Personal Workstations, ACM Press, New York, Addison-Wesley, Reading Mass., 1988
[10] ACM Oral History Interviews, ACM, New York, NY, USA, 2006.
ACM published 25 historic interviews.
[11] George Dyson, Turing's Cathedral – The Origins of the Digital Universe, Penguin Press Science, 2013,
[12] URL: liinwww.ira.uka.de/bibliography (access July 5, 2013)
[13] Kenneth O. May, Historiography: A Perspective for Computer Scientists, in: N. Metropolis et al, A History of Computing in the Twentieth Century, Academic Press, New York, 1980, page 16
[14] Jan-Willem Liebezeit, 2004, http://www.fmi.uni-jena.de/Fakult%C3%A4t/Institute+und+Abteilungen/Abteilung+f%C3%BCr+Didaktik/Didaktik+der+Informatik/Leibniz_Rechenmaschinen.html, access 03.04.2013
[15] Jeremy Avigad, John Harison, Formally Verified Mathematics, Communications of the AC, April 2014, Vol. 57, No. 4
[16] William Aspray, John von Neumann and the Origins of Modern Computing, The MIT Press, Cambrigde, Mass., 1990
[17] Norman Macrae, John von Neumann, Pantheon Books, New York, 1992

.

[18] Comparison of tables mostly from James W. Cortada, The Computer in the United States From Laboratory to Market, 1930 to 1960, M.E. Sharpe, Armonk, 1993

[19] Thomas J. Watson Jr., Peter Petre, Father, Son & Co. – My Life at IBM and Beyond, Bantam Books, 1990, page 377

[20] Whether binary digits and Boolean Logic are discoveries or inventions shall not be decided here. We use the word "invention" as we think that a very handsome notation was invented. A less comfortable notation to denote the same reality might have slowed down the process of computerization.

[21] J.A.N. Lee, Computer Pioneers, IEEE Computer Society Press, Los Alamitos, California, 1995

[22] Dennis E. Shasha, Cathy A. Lazere, Out of Their Minds: The lives and discoveries of 15 great computer scientists, Springer Verlag New York, 1998 (e.g. John Backus, John McCarthy, Alan C. Kay, Edgar W. Dijkstra, Michael O. Rabin, Donald E. Knuth, Edward A. Feigenbaum)

[23] Anthony Hyman, Charles Babbage Pioneer of the Computer, Princeton University Press, Reprint 1983

[24] Zuse, Konrad, Der Computer Mein Lebenswerk, Springer Verlag Berlin, 1984 English version: The Computer – My Life, Springer Verlag, 1993

[25] Allen B. Tucker, Peter Wegner, Computer Science: The Discipline and its Impact, in: Alan B. Tucker, Computer Science Handbook, Chapman and Hall / CRC, 2004, page 1-5

[26] http://www.ti.com/corp/docs/kilbyctr/jackbuilt.shtml (May 1, 2013)

[27] Edgar Frank Codd,

(1) A Relational Model of Data for Large Shared Data Banks, Comm. ACM, Vol. 13, 1970, No. 6;
(2) A Relational Model for Database Management: Version 2, Addison – WesleyPubl. Company, Reading Mass., 1990;

Cited according J.A.N. Lee, Computer Pioneers, IEEE Computer Society Press, Los Alamitos, Calif., 1995

[28] C.A.R. Hoare, C.B. Jones, Essays in Computing Science, Prentice Hall, 1989

[29] Jürgen Gulbins, UNIX, Springer, Berlin, 1988

[30] www.businessinsider.com/larry-page-the-untold-story-2014-4?op=1, (access on April 26, 2014)

[31] Die WELT, April 26, 2013

[32] IBM, Kleine Chronik der IBM Deutschland 1910 – 1987, IBM Form D 12-0017-8 (11/88)

.

---

[33] Allen B. Tucker, Peter Wegner, Computer Science: The Discipline and its Impact, in: Alan B. Tucker, Computer Science Handbook, Chapman and Hall / CRC, 2004, page 1-1

[34] James W. Cortada, The Computer in the United States From Laboratory to Market, 1930 to 1960, M.E. Sharpe, Armonk, 1993, page 136

[35] Eric G. Swedin, David L. Fero, Computers – The Life Story of a Technology, The John Hopkins University Press, Baltimore, 2005

[36] IBM, Kleine Chronik der IBM Deutschland 1910 – 1987, IBM Form D 12-0017-8 (11/88), page 91

[37] Emerson W. Pugh, Building IBM Shaping an Industry and Its Technology, The MIT Press, Cambridge Mass., 1995, pages 298

[38] http://www-03.ibm.com/ibm/history/history/year_1974.html, accessed May 16, 2013

[39] http://www-03.ibm.com/ibm/history/history/year_1974.html, accessed May 16, 2013

[40] Allen B. Tucker, Peter Wegner, Computer Science: The Discipline and its Impact, in: Alan B. Tucker, Computer Science Handbook, Chapman and Hall / CRC, 2004, page 1-1

[41] IBM, Systems Handbook, July 1986-II, GE19-5234-11

[42] http://europa.eu/legislation_summaries/information_society/other_policies/l24118_en.htm, access 09.06.2013

[43] http://www.buergerkarte.at/konzept/securitylayer/spezifikation/aktuell/introduction/Introduction.html, access 09.06.2013

[44] International Telecommunication Union, ITU-T H.323 (11/96), page i, "ITU (International Telecommunication Union) is the United Nations Specialized Agency in the field of
telecommunications. The ITU Telecommunication Standardization Sector (ITU-T) is a permanent organ of the ITU. The ITU-T is responsible for studying technical, operating and tariff questions and issuing Recommendations on them with a view to standardizing telecommunications on a worldwide basis." (from the same document, page ii)

[45] http://inspire.jrc.ec.europa.eu/ (access 09.06.2013)

[46] ACM: Association for Computing Machinery; www.acm.org

[47] Alan Clements, The Principles of Computer Hardware, 3rd edition, Oxford University Press, 2000

[48] C.A.R. Hoare, C.B. Jones, Essays in Computing Science, Prentice Hall, 1989, page 19ff

.

[49] Donald E. Knuth, The Art of Computer Programming, 2nd edition, Addison-Wesley, 1973 (1968)

[50] Emerson W. Pugh, Building IBM Shaping an Industry and Its Technology, The MIT Press, Cambridge Mass., 1995, pages 296ff, 307ff

[51] For more information: see Rudolf Güntsch in chapter 2.

[52] Thomas J. Watson Jr., Peter Petre, Father, Son & Co. – My Life at IBM and Beyond, Bantam Books, 1990, page 405

[53] Thomas J. Watson Jr., Peter Petre, Father, Son & Co. – My Life at IBM and Beyond, Bantam Books, 1990, page 401, 406

[54] Elliot I. Organick, The Multics System: An Examination of its Structure, MIT Press Cambridge Mass., 1972

[55] Thomas J. Watson Jr., Peter Petre, Father, Son & Co. – My Life at IBM and Beyond, Bantam Books, 1990, pages 372, 377

[56] IBM, IBM System /360 Principles of Operation, IBM Form A22-6821-1, June 1964

[57] Lawrence J. Kenah, Simon F. Bate, VAX/VMS Internals and Data Stuctures, Digital Equipment Corporation Press, 1984

[58] Hewlett-Packard Journal, August 1985, page 3

[59] Hewlett-Packard Journal September 1987, page 4 gives a complete introduction to the VLSI Processors for HP PA

[60] Hewlett-Packard Journal, August 1986, page 8

[61] Alan Clements, The Principles of Computer Hardware, 3rd edition, Oxford University Press, 2000, pages 353, 355

[62] www.arm.com

[63] All information of this paragraph is from the paper of Richard Comerford, How DEC developed Alpha, IEEE spectrum July 1992.

[64] IBM, Kleine Chronik der IBM Deutschland 1910 – 1987, IBM Form D 12-0017-8 (11/88), pages 86, 90

[65] E. Saab et al., Data Modeling and Design of a Patient Data Management System in an Intensive Care Unit, in: Computer Based Medical Systems, Proceedings of the Fifth Annual IEEE Symposium, June 14-17, 1992, IEEE Computer Society Press, 1992, page 54ff

[66] WW. Chu et al, A Temporal Evolutionary Object-Oriented Data Model for Medical Image Management, in: Computer Based Medical Systems, Proceedings of the Fifth Annual IEEE Symposium, June 14-17, 1992, IEEE Computer Society Press, 1992, page 84ff

[67] SAP, SAP HANA Administration Guide, SAP HANA Appliance Software SPS 05 Document Version: 1.0 - 2013-04-04

.

[68] Gerard J. Holzmann, Björn Pehrson, The Early History of Data Networks, IEEE Computer Society Press, 1995

[69] SkyDrive is a solution of Microsoft.

[70] http://www.theregister.co.uk/2013/01/17/google_orange/ (acces June 18, 2013)

[71] C.A.R. Hoare, C.B. Jones, Essays in Computing Science, Prentice Hall, 1989, page 3

[72] See e.g. www.chomskyinfo.info , accessed on May 20, 2013

[73] Christian Kirsch, Das Ende der Idylle, iX 4, 2006

[74] O.J. Dahl, E.W. Dijkstra, C.A.R. Hoare, Structured Programming, Academic Press, Lonon and New York, 1972 (available in ACM's digital library)

[75] Software AG, Introduction to Predict, Version 4.6.1, 2010

[76] Jean-Dominique Warnier, Logical Construction of Systems, John Wiley & Sons, New York, USA, 1981

[77] Michael Jackson, JSP in perspective, published in: Manfred Broy, Ernst Dehnert, Software Pioneers: contributions to software engineering, Springer Verlag, New York, 2002, pages 480-493

[78] Kevin M. Whitney, Fourth generation data management systems, US National Computer Conference, 1973

[79] Anthony Ralston, Chester L. Meek, Encyclopedia of Computer Science, Petrocelli Charter New York, page 350

[80] C. J. Date, An Introduction to Data Base Systems, Addison-Wesley, 1990

[81] Georg Schäfer, Mythen, BoD Norderstedt, 2006

[82] Mary E.S. Loomis, The 78 CODASYL Database Model: A Comparison with Preceding Specifications, ACM 0-89791-018-4/80/0500/003, 1980

[83] Günther Stürner, Oracle 7, dbms publishing Weissach, 1993

[84] Richard Barker, CASE*METHOD™ Entity Relationship Modelling, Addison-Wesley, 1990

[85] National Institute of Standards and Technology (NIST), http://www.nist.gov/manuscript-publication-search.cfm?pub_id=909505

[86] Bundesministerium des Innern, Europäisches Beschaffungshandbuch für offene Systeme – EPHOS ®, Glossar Version 1.0 de, 1995, ISSN 0179-7263

[87] http://www.cio.bund.de/DE/Architekturen-und-Standards/SAGA/saga_node.html, access June 10, 2013

[88] www.kepner-tregoe.com

[89] http://www.iso.org/iso/home/news_index/news_archive/news.htm?refid=Ref1662, accessed on May 20, 2013

.

---

[90] Patrick Thibodeau, What China's supercomputing push means for the U.S., Interview with Peter Beckman, computer scientist at the U.S. Department of Energy's Argonne National Laboratory and head of the DOE's exascale initiative, Computerworld June 10, 2013, 06:00 AM ET, ACM TechNews from June 12, 2013.

[91] Lifestyle and working habits of Seymour Cray and other pioneers of supercomputing are described by Charles J. Muray, The Supermen – The Story of Seymour Cray and the Technical Wizards behind the Supercomputer, John Wiley & Sons, Inc, 1997

[92] Thomas J. Watson Jr, Peter Petre, Father, Son & Co. – My Life at IBM and Beyond, Bantam Books, 1990, pages 407-410

[93] IBM Systems and Technology, Whitepaper, Watson – Antworten mit System, February 2011, POW03061-DEDE-00

[94] http://www.lrz.de/services/compute/supermuc/systemdescription/, access on June 13, 2013

[95] https://www.datacenter-insider.de/isc-2020-digital-der-erste-exascale-computer-a-941909/, https://www.fujitsu.com/emeia/about/resources/news/press-releases/2019/emeai-20190813-fujitsu-paves-the-way-to-exascale-computing.html, (September 3, 2020)

[96] NIST special publication 500-292, NIST Cloud Computing Reference Architecture, Sept. 2011

[97] NIST special publication 800-145, The NIST Definition of Cloud Computing, Sept. 2011

[98] Jerome Lobel, The cost of computer privacy, Honeywell Information Systems, Inc., Phoenix, Arizona, Paper published at the National Computer Conference 1974 (from ACM Digital Library, access May 26, 2013)

[99] Simitis (Hrsg.), Bundesdatenschutzgesetz, 7. Auflage, Nomos Verlag, 2011

[100] Frederick Lane, American privacy: the 400 – year history of our most contested right, Beacon Press Books, 2009

[101] Anthony Ralston, Edwin D. Reilly, David Hemmendinger, Encyclopedia of Computer Science, 4th Edition, ©2000,. John Wiley & Sons, Inc (ACM Digital Library, June 6, 2013)

[102] Directive 97/66/EC of the European Parliament and of the Council of 15 December 1997 concerning the processing of personal data and the protection of privacy in the telecommunications sector,

.

Directive 2002/58/EC of the European Parliament and of the Council of 12 July 2002
concerning the processing of personal data and the protection of privacy in the electronic communications sector (Directive on privacy and electronic communications),
(available from www.europe.eu)

[103] John Twelve Hawks, The Traveler, Doubleday Broadway Publishing Group, Random House Inc., New York, 2005

[104] Chris Edwards, Using Patient Data for Personalized Cancer Treatments, Communications of the ACM, April 2014, Vol. 57, No. 4, p. 13

[105] Schmidt, Eric, Cohen, Jared, The New Digital Age, John Murray Publishers, London, 2013

[106] Stas Filshtinskiy, Privacy and Security – Cybercrime, Cyberweapons, Cyber Wars: Is There Too Much of It in the Air?, Communications of the ACM, June 2013, Page 28

[107] „Orientierungshilfe – Cloud Computing" der Arbeitskreise Technik und Medien
der Konferenz der Datenschutzbeauftragten des Bundes und der Länder Version 1.0
Stand 26.09.2011, page 14

[108] Jean-Louis Bravard, Robert Morgan, Smarter Outsourcing, Prentice Hall, 2006, page 1 of the Foreword

[109] Jean-Louis Bravard, Robert Morgan, Smarter Outsourcing, Prentice Hall, 2006, page 2

[110] http://business.chip.de/news/OpenSSL-Heartbleed-So-sichern-Sie-Ihren-Rechner-ab_69059617.html (access April 12, 2014)

[111] The German Agency for Public Procurement publishes such a handbook about public procurement which is helpful for private and public organizations. Beschaffungsamt des Bundesministeriums des Innern Deutschland, Unterlage fü Ausschreibung und Bewertung von IT-Leistungen, Version 2.0, Apil 2014, www.beschaffungsamt.de

[112] http://www.data.gov

[113] Georg Schäfer, Bürger wollen eine Regierung zum Mit-Machen – Open Data und Open Government, BoD, 2012

[114] DLA Piper, EU study on the legal analysis of a single market for the information society: new rules for a new age?, Brussels, Nov. 2009 (in response to the invitation to tender OJ 2007 / S 202 244659 of 19/10/2007)

.

[115] Bizer, C., Heath, T., Berners-Lee, T.: Linked data – the story so far. International Journal On Semantic Web and Information Systems 5(3), 2009, pages 1-22
[116] www.sunlightfoundation.com
[117] http://liquidfeedback.org/ (access June 17, 2013)
[118] Stephen B. Wicker, Stephanie M. Santoso, Viewpoint – Access tot he Internet Is a Human Right, Commnications of the ACM, June 2013, page 43
[119] Van Schewick, Barbara, Internet Architecture and Innovation, 2010, ISBN 978-0-262-01397-0

[120] Die Welt, Das endlose Geranel ums freie Internet, March 13, 2014
[121] http://ec.europa.eu/digital-agenda/en/eu-actions (accessed on April 6, 2014)